The Extraordinary Life of Josef Ganz

Paul Schilperoord

THE EXTRAORDINARY LIFE OF

JOSEF GANZ

THE JEWISH ENGINEER
BEHIND HITLER'S

VOLKSWAGEN

RVP PUBLISHERS
NEW YORK

For Josef Ganz

RVP Publishers Inc.
95 Morton Street, Ground Floor
New York, NY 10014

First published in 2012 by RVP Publishers.

Second edition

The Extraordinary Life of Josef Ganz: The Jewish Engineer Behind Hitler's Volkswagen is an extended edition of *Het ware verhaal van de Kever,* published in the Netherlands in 2009 by Uitgeverij Veen Magazines, Diemen.

Translated from the Dutch by Liz Waters

This book was published with the support of the Dutch Foundation for Literature.

Designed by Bruno Herfst

Library of Congress Control Number: 2011945771

ISBN 978-1-61412-203-6

www.rvpp.com

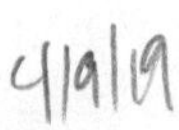

CONTENTS

INTRODUCTION

The Volkswagen Beetle is the most popular car of all time. Over a period of sixty-five years, more than 21.5 million were built, and the adorable automobile grew into a symbol of the peace-loving hippie movement—in sharp contrast to its origins as the People's Car of Nazi Germany. It is common knowledge that Adolf Hitler commissioned Ferdinand Porsche to develop the Volkswagen with state funding, but hardly anyone is aware that Hitler poached the Volkswagen idea and Porsche the technical concept from a Jewish engineer, Josef Ganz. It was he who created the basic design, insisted year after year that an affordable car should be produced for the ordinary citizen, and as early as 1931, drove around Frankfurt in a successful prototype nicknamed the May Bug before eventually, in Hitler's presence, introducing a serial-production version as the "German Volkswagen."

Ganz and Hitler both rose to prominence at a time of worldwide financial calamity. Josef Ganz promoted innovative cars as an alternative to the fleet of large, heavy, and—especially after the stock-market crash—unaffordable automobiles produced by established manufacturers, while Adolf Hitler gave Germans fresh hope by finding a scapegoat for the economic crisis. The fact that the Volkswagen concept was developed by a Jew did not deter Hitler from making it his own, deliberately erasing Josef Ganz's name from the history books in the process. Porsche, by contrast, will always be remembered as a world-famous car designer. Partly because of the public eulogies and official honors showered on him by Hitler, he gained a mythical status that endures to this day—as does the close connection between the Porsche and Volkswagen brands, reinforced by the merger of the two carmakers in 2009 at the initiative of Ferdinand Porsche's grandson Ferdinand Piëch, who now serves on the supervisory board of both companies.

Despite an intense interest since childhood in the Volkswagen Beetle, I had never heard of Josef Ganz before coming across his name in a 1980 issue of the American car magazine *Automobile Quarterly* that I bought over the Internet in 2004. There I read that Josef Ganz was the forerunner to Ferdinand Porsche and that he had developed a Volkswagen years earlier, with the typical Beetle characteristics: rear-mounted engine, backbone chassis, swinging rear half-axles, and streamlining. Not only had he built several such models, but he had consistently advocated their introduction from 1928 onward, as editor-in-chief of the progressive trade magazine *Motor-Kritik*. In 1933, Ganz was arrested by the Gestapo and disappeared from the German stage.

Intrigued by these sparse facts, I decided to investigate Josef Ganz's role in the development of the VW Beetle. There was just one concrete starting point for my research: a full set of back issues of *Motor-Kritik* in the library of the Deutsches Museum in Munich. I flew to Munich and spent two days searching through the magazine, finding several articles that supported the claims of *Automobile Quarterly*, with detailed descriptions of Ganz's revolutionary prototypes, photographs of them, and references to his arrest by the Gestapo. Within a year of that arrest, the influential editor-in-chief seemed to have vanished without trace. This initial research fired my imagination: why did we know so little about a man who, during the time he was employed as technical consultant to major companies including BMW and Mercedes-Benz, worked on revolutionary models? After my visit to Munich, I started writing to German and Swiss museums and archivists. This rendered up a number of interesting documents, sources of information, and clues that I investigated further. Slowly I managed to piece together the general outlines of Josef Ganz's story.

In January 2005, I published the early results of my research in a cover story for the Dutch technology magazine *De Ingenieur*, and not long afterwards, an article on the subject appeared in the Swiss daily newspaper *Tages Anzeiger*. I was then approached by one Dieter Klüpfel, a Canadian resident who had been sent the newspaper article by a Swiss friend. He turned out to be none other than the nephew of Josef Ganz's long-term girlfriend, and he told me that he had lived for many years in the couple's house. In a number of telephone conversations and countless e-mails, he described what he could remember and later gave me a photograph album with dozens of pictures of Ganz and his prototypes. Later that year, I met up with Dieter Klüpfel in Switzerland, an encounter that was filmed for the Dutch television program *Netwerk* as part of a report on my investigation. We also visited the Volkswagen company archives together, where there was a rather muted response to questions about the part played by Josef Ganz. After the publication of a follow-up article in the German magazine *Technology Review*, Volkswagen became a little more willing to assist in my research and made a small amount of archive material available. As I studied it and other documents that came to light, the idea of developing the story into a book began to take shape.

Another interesting contact was Gerry Harant, an old friend of "Joe" Ganz in Australia, whom I met over the Internet. He too told me everything he could remember. He managed to track down Ganz's last will and testament, which stated that he wished to leave his archive to a Swiss relative, who turned out never to have received it. I continued searching for the archive and traced the family of Ganz's then lawyer, who to my amazement and delight had kept a large number of photographic negatives and microfilms for several decades, knowing that Ganz had been "very important to the Volkswagen." This was a source of invaluable material, including testimony about a Gestapo officer who had hounded Ganz, as well as photographs of test drives in prototypes built by Mercedes-Benz and correspondence with the directors—items unavailable even in the Mercedes-Benz archives. In fact, company records at BMW and Mercedes-Benz contain no information about Ganz at all. The photographic archive from Australia was followed by another important discovery, a file about Ganz left by a Swiss journalist. These new documents contained crucial pieces of the puzzle. I began to try to reconstruct Josef Ganz's highly complex and turbulent life.

On completing this extensive preparatory work, I decided to write Ganz's story in narrative form. The many hundreds of letters, articles, personal statements, legal documents, patents, and photographs available to me served both as supporting evidence and as a means of adding atmosphere to my account.

It remains an incontestable fact that in designing and building the Volkswagen Beetle, Ferdinand Porsche and his engineering team created an astonishingly sound and reliable car for its time. But Porsche was not the originator of the concept. It is no exaggeration to say that the immensely popular VW Beetle would never have existed without Josef Ganz. He was a central figure in the pre-war German car industry, and an account of his life makes the ideal framework for a book about developments that ultimately produced the Volkswagen. Ganz summed up his own thrilling tale perfectly a few years before his death when he wrote: "I promise you an insight into a crime story even Hitchcock could not invent."

Paul Schilperoord

PROLOGUE

In the early morning of February 11, 1933, Berlin was covered in a deep layer of fresh snow. The streets were treacherously icy, and the slow crawl of traffic contrasted markedly with recent tempestuous political developments in the German capital. For twelve days, a country suffering economic depression had been ruled by a new regime, led by the fascist Nazi Party. From their hotel rooms, two men watched the white flakes gently drift down out of a dark, threatening sky onto the creeping procession of automobiles, streetcars, cyclists, and pedestrians. For both men, the spectacle provided a brief peaceful interlude in an otherwise turbulent time. Today was to be no less momentous. It was the day of the official opening of the International Automobile and Motorcycle Exhibition, the IAMA, the first motor show to be held in Berlin for two years. It had unexpectedly acquired a powerful political charge. This major international event would be a moment of glory for both Adolf Hitler and Josef Ganz.

Hitler had been sworn in as German chancellor by President Paul von Hindenburg on January 30, leading a coalition government composed of his own Nationalsozialistische Deutsche Arbeiterpartei (the National Socialist German Workers' Party, or Nazi Party) and the Deutschnationale Volkspartei (DNVP). Since serving a year in prison after a failed coup attempt in late 1923, Hitler had been determined to take power by legal means. Now, less than ten years later, he had succeeded. Although the Nazi Party had been insignificant in size during the economically better times of the late 1920s, as a result of the Depression Hitler's promises of a new economically and politically powerful Germany had won him a huge following. The recently appointed chancellor seized the approaching IAMA with both hands, preparing to open the show with a bombastic speech in which he would present himself as a decisive leader and statesman. That speech, which now lay neatly typed on his bedside table, announced the first of his government's plans for the motorization of Germany. Although he had never possessed a driver's license, Hitler was a fervent fan of the automobile and a strong advocate of mobility for the masses. In early February, the Ministry of Transportation had announced that small four-wheeled cars could from now on be driven by anyone holding a license to ride a motorcycle. With that decision, a few days after taking power, Hitler introduced a measure for which Josef Ganz had been campaigning persistently for years.

For Ganz, nine years younger than Hitler, the 1933 IAMA represented the crowning achievement of five years of hard work. As an independent engineer and editor-in-chief of the much-praised but highly controversial trade magazine *Motor-Kritik*, he set himself the goal of revolutionizing the stagnant, old-fashioned German car industry. Throughout those five years, he attacked German manufacturers in his critical articles, describing their cars as far too expensive, ungainly, heavy, unwieldy, unsafe, and outdated. By becoming the first technically brilliant journalist to express such criticism openly, he had made enemies within the industry—highly placed enemies who tried repeatedly to undermine his growing influence. They publicly accused him of blackmailing manufacturers and of sabotaging the German economy and dismissed his magazine as a sensation-seeking rag. That "rag"—*Motor-Kritik*—was in fact a leading journal of the industry, and it was attracting more readers by the day. Ganz had established conclusively that cars could be made lighter and safer, more reliable and more efficient. His ideas were freely available, but instead of turning them to their own advantage, car manufacturers attempted to ruin him. Even after his opponents organized an advertising boycott of his magazine, instituted legal proceedings to silence him, and mounted smear campaigns to wreck his reputation, Josef Ganz's progressive spirit won out against technical conformism. Now at last manufacturers were showing new models at the IAMA built according to Ganz's guidelines—the first ever lightweight cars with fully independent suspension, optimal driving qualities, and early signs of streamlining. Ganz's most important showpiece was a Volkswagen built to his own design, with the engine at the back. The concept was so revolutionary that over the previous few years other manufacturers had tried to steal the design and obstruct the building of his prototypes.

The Berlin motor show Hitler was due to open that day manifestly symbolized the revolution that was taking place in Germany, both in politics and in the car industry. The political revolution had been brought about by a self-declared Aryan, Adolf Hitler, and the industrial revolution by Josef Ganz, a Jew. Both men would be present at the IAMA, where their ideas about the motorization of Germany would come together, combine—and change hands. Hitler would stand eye to eye with Ganz's Volkswagen, the cheapest four-wheeled car in Germany, which because of his new legislation could be driven by anyone with a motorcycle license. This combination of a technically advanced People's Car and legislation to suit it was so spectacular that it received worldwide coverage. A journalist for *The Detroit News* concluded: "Whatever may be the future development of this type of car in Germany, without doubt the Hitler government will be responsible for its popularization."[1] Although Hitler would indeed popularize the Volkswagen, even the most prophetic journalist could not have foreseen that in the process the statesman would deliberately erase the brilliant Jewish engineer Josef Ganz from history and hand his brainchild over to Ferdinand Porsche.

CHAPTER ONE

FROM TROPFENWAGEN TO VOLKSWAGEN

Dr. Hugo Markus Ganz had been living in Budapest for some time when his son Josef was born on a summery first of July 1898. Ten years earlier, the German daily *Frankfurter Zeitung* had posted him to the Austro-Hungarian double monarchy, an ally of the German Empire, as its foreign correspondent. Dr. Ganz came originally from a Jewish family in the German city of Mainz and had studied history and philosophy in Leipzig, but he was starting to feel very much at home in Budapest. It was there that he met his wife, the Hungarian-Jewish Miriam Török, a progressive woman from a well-to-do family. Her brother was a famous dermatologist and a university professor, and her uncle was the former president of the Hungarian royal supreme court. Their first child was born in Budapest in September 1893, a daughter called Margit. Now, five years later, she had a younger brother.

Shortly after the birth of his son, Hugo Ganz's work took him and his family to Vienna, the capital of the Austro-Hungarian Empire and the cultural and intellectual center of Europe. The city was in its heyday at the turn of the century, attracting eminent figures, including famous composers, philosophers, psychiatrists, and artists. For the journalist, writer, and diplomat Hugo Ganz, this was an extremely stimulating environment. It was here that he wrote *Der Rebell*, a play in five acts that was published in 1900, followed two years later by a book called *Zur Kunstrede des deutschen Kaisers*, a critical commentary on a famous speech given by the German Emperor Wilhelm II in late 1901, in which the Kaiser turned his back on modern art that did not suit his personal taste or his legislative program. For Hugo's children Margit and Josef, who regularly went into town with their mother, busy Vienna, with its bustling streets and massive, imposing buildings, was magical. In contrast to his father, a man so well versed in literature, Josef was a born technician. As a young child, he was fascinated by the great technological changes underway in Europe at the beginning of the new century. In Budapest, his mother had taken her baby son for rides in the electrically powered metro, the first subway network in continental Europe, which had opened two years before he was born.

Birthplace of Auto Engineering

In busy Vienna, Josef Ganz feasted his eyes. He saw the last horse-drawn streetcars taken out of service to be replaced by steam and the first electric streetcar lines. Even more significantly, Austria-Hungary was the birthplace of auto engineering, where in 1870 the inventor and automobile pioneer Siegfried Marcus had built the world's first vehicle to be powered by an internal combustion engine.[1] Like Josef Ganz, Siegfried Marcus was a technical genius of Jewish extraction. Symbolically perhaps, he died in Vienna on July 1, 1898, the day Josef was born. After Marcus's early experiments, another twenty years would pass before motorized transportation began to come into its own around 1890, with German pioneers such as Karl Friedrich Benz of Benz & Cie and Gottlieb Daimler of Daimler Motoren Gesellschaft (DMG) putting their first cars into serial production.

The vehicles of those years were still without exception motorized coaches, high on their large wagon-wheels, but in Josef Ganz's early youth the automobile broke with tradition and changed its appearance radically, adopting a basic shape of its own that was modeled after the 1901 Mercedes 35 HP built by DMG. The engine was moved from the back to the front, under a hood with a large upright radiator ahead of it, though it still drove the rear wheels. The chassis was no longer made of wood but of pressed steel. It was around this time that a new generation of trailblazing Austrian engineers, Ferdinand Porsche, Edmund Rumpler, and Hans Ledwinka among them, built their first cars. Many of those early models could be seen in the streets of fin de siècle Vienna as Josef Ganz was taking his first steps. He later enjoyed identifying the different makes of big, imposing automobiles by their gleaming radiators: Benz, Daimler, Laurin & Klement, Lohner, Nesseldorfer Waggonbau (NW), Panhard & Levassor, and Puch. The technology young Josef encountered in the streets of 1900s Vienna was relatively primitive, and while still a young child he began to see all kinds of room for improvement. At the age of nine, he applied for his first patent, for a safety system for the electric streetcar that would later find widespread application.[2]

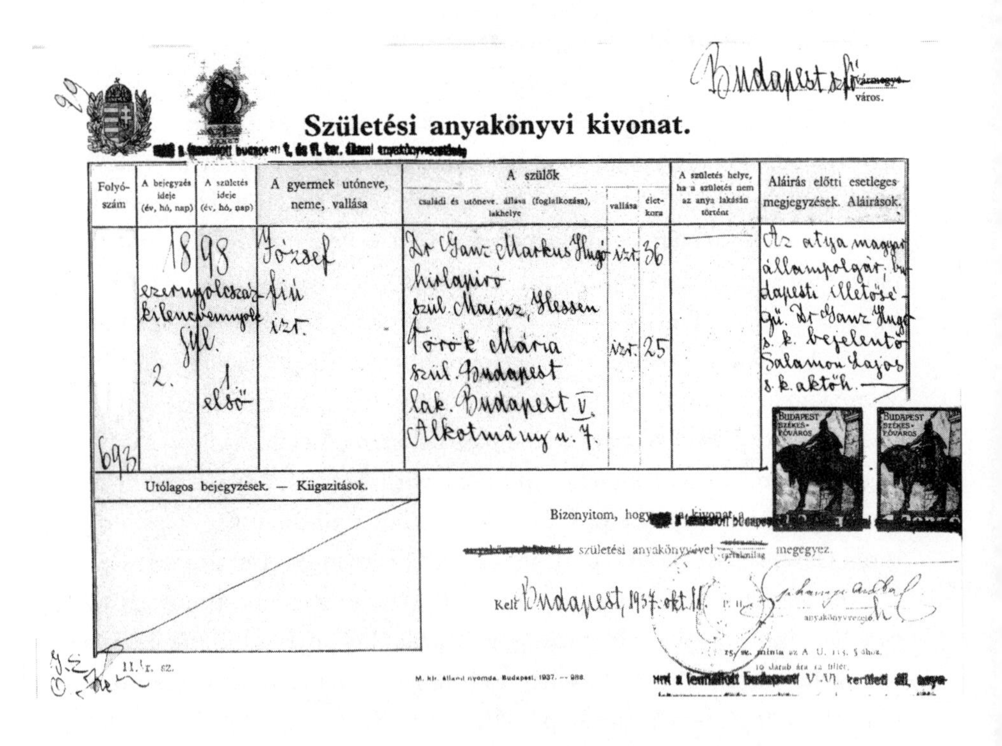

Budapest szfő város.

Születési anyakönyvi kivonat.

Folyószám	A bejegyzés ideje (év, hó, nap)	A születés ideje (év, hó, nap)	A gyermek utóneve, neme, vallása	A szülők: családi és utóneve, állása (foglalkozása), lakhelye	vallása	életkora	A születés helye, ha a születés nem az anya lakásán történt	Aláírás előtti esetleges megjegyzések. Aláírások.
693	1898 ezernyolcszáz kilencvennyolc júl. 2.	98 1 első	József fiú izr.	Dr Ganz Markus Hugó hírlapíró szül. Mainz, Hessen	izr.	36	—	Az atya magyar állampolgár, budapesti illetőségű. Dr Ganz Hugó s. k. bejelentő Salamon Lajos s. k. aktőh.
				Török Mária szül. Budapest lak. Budapest V. Alkotmány u. 7.	izr.	25		

Utólagos bejegyzések. — Kiigazitások.

Bizonyitom, hogy a kivonat a … születési anyakönyvvel … megegyez.

Kelt Budapest, 1937. okt. 11.

P. H.

anyakönyvvezető.

11. r. sz.

M. kir. állami nyomda. Budapest, 1937. — 988.

Josef Ganz's birth certificate, Budapest 1898

Parties in Vienna

Sometimes, months might go by in which Margit and Josef barely saw their father. His time and attention were swallowed up not just by everyday journalistic work but by the long trips he took. For his *Reiseskizzen aus Rumänien* (*Travel Sketches from Romania*), he traveled all over that country in 1903, and in early 1904 he was stationed in St. Petersburg for the first three months of the Russo-Japanese War. In the well-received book he wrote there, *Vor der Katastrophe* (*The Land of Riddles: Russia of Today*), which was translated into many languages, he described the profoundly divided Russia of Tsar Nicholas II. As a universally respected journalist and diplomat, Hugo Ganz had access to Europe's dignitaries, from ministers to kings. In Vienna, Hugo Ganz was appointed President of the Foreign Press Association, and his books and articles brought him increasing fame and respect. As well as the *Frankfurter Zeitung*, he wrote for the Swiss newspaper *Neuen Zürcher Zeitung*, for which, as part of a small Jewish minority, he covered among other things issues affecting Europe's Jews.

Hugo Ganz garnered so many friends and acquaintances as an influential commentator that he and his wife Miriam, who preferred to be called Maria, turned their grand family home in Vienna into a meeting place for famous statesmen, scientists, writers, actors, and artists from all over the world.[3] Hugo was a friend of the philosopher Tomáš Garrigue Masaryk, for example, later to become the first president of Czechoslovakia; of the Austrian writer and translator Siegfried Trebitsch; and of the Austrian painter Franz von Bayros, who, as a devotee of the Decadent movement, was known primarily for the erotic subject matter of his art. Hugo Ganz was able to appreciate the artistic value of such works, but writing in the *Frankfurter Zeitung* in 1910, he opposed the public exhibition in Vienna of the erotic works of Gustav Klimt, for the sake of "protecting bourgeois morals." He wielded considerable influence. The career of the Austrian writer Karl Schönherr, another of his friends, received an enormous boost when Hugo Ganz wrote about him on several occasions for the *Frankfurter Zeitung*.

When his patent for a streetcar safety system was granted, twelve year-old Josef was himself featured in the magazine *Stadt Gottes*. The short article said that "the young technician already has made multiple remarkable inventions" and was destined to become "a famous man."[4]

For little Josef and his sister Margit, this was a fascinating and inspiring environment in which to grow up. At an early age, they came into contact with an astonishingly diverse range of people and with progressive ideas that went far beyond the commonly accepted norms and values of the day, which were steeped in religion.

During the opulent parties held regularly by their parents in the vast rooms of their house, Josef and Margit often crept out of bed. They were on good terms with the young Czech composer and pianist Erich Wolfgang Korngold, who often performed at their home and composed many of his new works at the family piano. He was extremely taken with Margit Ganz, and every year he wrote a special piece for her birthday. During parties, Josef and Margit would hide at his feet under the big Steinway grand, watching with admiration as politicians and eccentric artists engaged in heated debates. Besides the house pianist Korngold, there were regular performances by the famous Rosé Quartet, led by the Jewish violinist Arnold Rosé, who worked closely with composer Johannes Brahms and married Gustav Mahler's sister Justina. Hugo Ganz was a good friend of Arnold Rosé and his fellow musicians.[5] On those festive evenings, Josef hugely enjoyed listening to the quartet, which played mainly works by Haydn, Mozart, and Beethoven. His technical aptitude aside, Josef was a talented musician who could play the violin beautifully. His sister Margit also had an artistic flair, but in her case it was manifested in the theater. She became so famous for her acting talent that when she was twenty a bust was made of her by the profoundly deaf Austrian sculptor Gustinus Ambrosi, one of a group of young artists who regularly visited the Ganz household. He was the same age as Margit Ganz, just twenty in 1913 when Emperor Franz Joseph I, recognizing his great talent, allocated him a studio at state expense. There he made busts, including those of famous politicians, physicians, industrialists, composers, and artists. On the eve of the First World War, he immortalized not only Margit Ganz but other artistic figures, including composer Richard Strauss and the author Stefan Zweig.

First World War

In the spring of 1914, Margit Ganz fell in love with the Jewish actor Jakob Feldhammer, eleven years her senior. He had been invited to play the lead in Shakespeare's *Hamlet* at Vienna's Neuen Wiener Bühne, and Margit was cast as Ophelia. Dr. Emil Geyer, the director of the theater and a friend of Hugo Ganz, had introduced the couple, and he asked Feldhammer to prepare Margit Ganz for her role. As a result, Feldhammer became a regular visitor to the Ganz household and got to know the actress's younger brother Josef. After signing his contract with the Vienna theater, however, Feldhammer left for a summer holiday with his sister Anna and a couple with whom they were close. They traveled by train to Florence in Italy, where the two women stayed while Jakob and his friend took a long trip around the country in an open sports car. Jakob Feldhammer returned from his travels just before Josef's sixteenth birthday. Josef thoroughly enjoyed listening to the charismatic Feldhammer's thrilling stories about his adventurous journey. Feldhammer and his friend had driven from Florence to Rome and Naples, then all the way south to Palermo in Sicily. There they bought new parts for the car and traveled on by ship to Tunisia before driving to Algeria and taking another ship to Marseille. Josef Ganz was enthralled by tales of how people in these exotic countries cheered from the roadside as they passed and how the two men took high-speed bets on which tire would blow first, off-side front or near-side rear. They had no fewer than six spare tires on board, which had to be levered onto the rim at the roadside. "It was always the one we least expected," Feldhammer told the fascinated Josef.[6]

The rest of the summer was overshadowed by events in Sarajevo, the capital of the Austro-Hungarian province of Bosnia-Herzegovina. On June 28, 1914, three days before Josef turned sixteen, the Austrian crown prince Franz Ferdinand, on an official visit to the city, was assassinated by the Bosnian-Serb student Gavrilo Princip. On Josef's birthday and for the days that followed, Hugo Ganz had his hands full reporting on the subject for the *Frankfurter Zeitung* and the *Neuen Zürcher Zeitung*. Several weeks of diplomatic negotiations followed before, at the end of July, first Austria-Hungary, then Russia and France, began to mobilize. The many pacts and alliances between European powers created a threat of hostilities on an unprecedented scale. When war broke out in early August 1914, the mood in Europe was naively optimistic. Eager young men, convinced the conflict would be over within weeks, rushed to volunteer. At sixteen, Josef Ganz was not yet old enough to enlist, but in a purely technical sense this war must have fascinated him. It brought the first large-scale deployment of inventions such as automatic weapons, tanks, and aircraft. Ganz made his own technical contribution by developing an automated aiming device for anti-aircraft guns. He was seventeen.[7]

In 1916, with the war that had begun amid such optimism bogged down in the Flemish mud, the Ganz family moved to Germany, Austria-Hungary's principal ally, where they took German citizenship. Josef accompanied his parents, but after finishing secondary school in Frankfurt that summer he decided to return to Vienna alone to study mechanical engineering at the Technical College. He would not attend many lectures. In the fall, four months after his eighteenth birthday, Josef Ganz, like so many of his contemporaries, enlisted as a volunteer. At his own request, he was sent to join the seaplane section of the German Imperial Navy. In March 1917, having completed basic training, he was deployed to the Second Naval Flying Squadron in Wilhelmshaven on the Western Front and in June the same year was transferred to the First Naval Flying Squadron in Warnemünde on the Eastern Front. There, as an ordinary seaman serving on an aircraft carrier, he was seriously wounded during heavy fighting. His groin and navel were ruptured and required surgery. Jakob Feldhammer, his sister's boyfriend, a lieutenant in the Royal and Imperial Army, was also wounded, and both men spent time convalescing at Josef's parental home in Vienna. On those cultural Sunday evenings, with performances by young artists and writers, the war seemed a world away.[8]

Although not yet fully recovered, Josef Ganz soon returned to the seaplane base in Warnemünde, where on March 28, 1918, he was promoted to the rank of *Flugzeug-Obermatrosen*. He became good friends with Dr. Michael Max Munk, eight years his senior, a pioneer of aerodynamics who supplemented Ganz's technical education. Munk offered him the opportunity to work on aircraft as an engineering technician, investigating ways of installing weapons and carrying out a wide range of scientific measurements. Ganz helped Munk to develop a wind tunnel and became fascinated by streamlining.[9] He was surprised that such research was carried out only on aircraft and zeppelins, since streamlining was also capable of making automobiles faster, more efficient, and safer.

The intensive technical activity at the Warnemünde airbase distracted young Josef from the gruesome everyday realities of the war. It seemed as if an entire generation of young men, in many cases several years younger even than he was, were paying for this senseless conflict with their lives or suffering horrible mutilation. Although seriously wounded, Ganz at least had the good fortune to survive. After more than four years of war, the German Empire and the Allies finally signed an armistice on November 11, 1918, bringing an end to the industrialized slaughter. All told, the World War had cost more than eight million lives and left twenty-one million people wounded, as well as reducing large areas of Europe to wasteland. When it was over, Josef Ganz continued to serve in Warnemünde for several weeks as a technical assistant to the designer Max Munk before their ways parted for good. With his considerable expertise, Munk was transferred to the United States by the National Advisory Committee on Aeronautics (NACA), the forerunner of the National Aeronautics and Space Administration (NASA). Among his contributions to its research was the development of a revolutionary type of wind tunnel.

Christmas in Vienna

Amid the chaos of postwar Europe, Josef Ganz decided to return to Vienna in late December 1918 to continue his studies at the Technical College. During the long train journey, he saw with his own eyes some of the devastation left by the fighting. The Vienna he arrived in was no longer the capital of the Austro-Hungarian dual monarchy but of Austria, an independent republic on a par with the former imperial territories that were now the nation states of Hungary, Czechoslovakia, and Yugoslavia.

In Vienna, he was reunited with his parents. His sister Margit had moved to Frankfurt, having married her great love, the actor Jakob Feldhammer, the previous year. Josef's wounds and the aftermath of war meant it was some time before he could fully concentrate on his studies again, but although he had been unable to attend many of his classes for that academic year, he left for Switzerland in late July 1919 armed with several certificates from the Technical College.

There he would spend the summer with his uncle, Alfred Ganz, a rich industrialist who lived in the beautiful Villa Solina in the village of St. Niklausen, near Luzern. Josef was fond of his uncle, his father's younger brother by twelve years, a man seen by the family as a rather eccentric character who had made his fortune manufacturing carbide lamps for use by miners and in car headlamps. To the exasperation of some of his relatives, Alfred spent his wealth freely, amassing a large collection of artworks, mainly statuettes and other objects made of tin as well as paintings by famous artists. In that immediate post-war period, Switzerland was a peaceful haven, while Germany was plagued by government crises, left-wing and right-wing uprisings, paramilitary violence by Free Corps made up of veterans who refused to lay down their arms, famine, and outbreaks of disease. After a few delightful summer months, Josef Ganz finally had to leave the Villa Solina in October 1919 and move to Worms in Germany, where two months later he began an apprenticeship at the chemicals factory Chemischen Fabriken Worms AG.

Debates on Auto Engineering

After less than two months, the directors of the factory were so happy with Josef's work as an apprentice that they offered him a post as factory engineer, beginning on February 1, 1920. Still only twenty-one, Josef Ganz was given astonishing freedom for someone of his age and experience to concentrate on designing and building a whole new factory workshop with a chemical laboratory attached.[10] With his new salary, he could finally get himself a motorcycle license. He had an appropriately stylish photograph taken for it, wearing a heavy, long, dark-brown leather motorcycling jacket. He even managed to put together enough money to buy a secondhand motorcycle. It was a Wanderer, a German brand, and Ganz maintained it himself. At one stage, he took the engine out of the frame and dismantled and reassembled it, so as to make a detailed study of the mechanism. He would have loved to buy a car, but such a luxury was beyond the means of a factory engineer. At the chemical works, he got to know Stephan Mittler, a fellow engineer with a comparable passion for cars and motorcycles. During lunch breaks and increasingly after work as well, they discussed auto engineering. During one conversation, Stephan raised the question of why car engines were mounted at the front. Surely the rear was more logical, he argued, since the engine drove the rear wheels, and a long drive shaft was a source of mechanical friction and vibration. Ganz initially said there must be extremely good reasons why auto engineers, after extensive experimentation in the pioneering days of the automobile, had unanimously opted for front-mounted engines. He joked: "Now, twenty years later, we're surely not going to start all over again, by marching backwards!"[11] Later that evening, stimulated by his discussion with Stephan, Josef Ganz set off home on his motorbike. As he struggled to find his way along the pitch-black, unlighted streets, Stephan's "idiotic" idea played on his mind.

"We were extremely happy with his performance,"[12] wrote the directors of the Chemischen Fabriken Worms AG on December 1, 1920, in a glowing reference. Having enjoyed his time in Worms, Josef Ganz had decided to move to Darmstadt, where he would continue his studies. Despite the close friendship that developed between him and his colleague Stephan Mittler, the two young men subsequently lost touch. Stephan moved to Nuremberg and later returned to Vienna. In Darmstadt, Josef found himself a room at Riedeselstrasse 25, in the house of the Paqué family, where he would meet his first great love, and enrolled at the Darmstadt Technical College to study mechanical engineering. When he began the new term in January 1921, his head was buzzing with ideas, and in the evenings, alone in his room, he created a wide range of technical designs. In September that year, he put in a patent application for a mechanism to drive a ship's screw,[13] but he concentrated above all on radio technology and auto engineering and began working on an idea for a small car with a rear-mounted engine. It was intended as a vehicle ordinary people could afford and as a safer alternative to the motorcycle, which had already caused Josef Ganz quite a few injuries in falls.

Drop Car

When the hulking black steam train, pouring smoke, finally hissed into Berlin's main station, twenty-three-year-old Josef Ganz could hardly contain his joy. He had been traveling for hours, all the way from Darmstadt, for the official opening on September 23, 1921, of the German Automobile Exhibition. He had been looking forward to it for a long time. This was the first motor show to be held in Berlin since the war and an unprecedented opportunity for Ganz to see all the most important makes of car in one place, including Germany's Adler, Benz, Daimler, and Opel, Austria's Austro-Daimler, Puch, and Steyr, and Czechoslovakia's Laurin & Klement and Tatra. Among all those similar automobiles, with their big gleaming radiators, was one model that was radically different, a car that would fascinate and inspire Josef Ganz endlessly. Its German manufacturer, Rumpler, had called it the Tropfenwagen, or "drop car," after the perfect streamlined shape of a falling water droplet. The engineer Edmund Rumpler had applied his knowledge of aircraft aerodynamics to the new model, which unlike all the other cars at the fair had a shape as slippery as an eel, complete with curved windows, a single headlamp at the center of the bodywork, and fenders in the form of large fins. The engine was mounted not at the front but in the rear, just ahead of the rear wheels and their independent suspension. Josef Ganz stood and stared at the impressive colossus, which looked like a giant fish on wheels. The image of the drop car would haunt him. He predicted a great future for passenger vehicles of this type: futuristic, streamlined models with efficient transmission and optimal road holding. The drop car, Ganz thought, might even become the model for that revolutionary, much smaller, universally affordable vehicle he had in mind: the "Volkswagen," or "People's Car."

Throughout the long journey back to Darmstadt, as the German landscape slipped by, Ganz contemplated his idea. It would be a huge challenge but one he was keen to take on. Nobody had ever succeeded in developing a simple car that rode well and had no major shortcomings yet could be sold at a price ordinary people had a chance of being able to pay. In France and Britain especially, far smaller models had been built using motorcycle engines, the so-called cyclecars, but they were relatively primitive. At the exhibition in Berlin, Ganz had seen nothing that came even close to what he had in mind, other than the far larger Rumpler. In early 1922, he sent a letter to the editors of the car and motorcycle magazine *Klein-Motor-Sport,* in which he expounded upon the dangers of riding a motorcycle on poor German roads, with surfaces made of "arched cat-spines in porridge." He said there was a pressing need for a "cheap four-wheeled vehicle" at the same purchase price and with the same maintenance costs as a good motorcycle. The article included a basic description of the kind of car he had in mind, with an air-cooled, four-stroke, rear-mounted "horizontally opposed" or "boxer" engine made of lightweight alloy, built as a single unit, with a clutch and three-speed gearbox.[14]

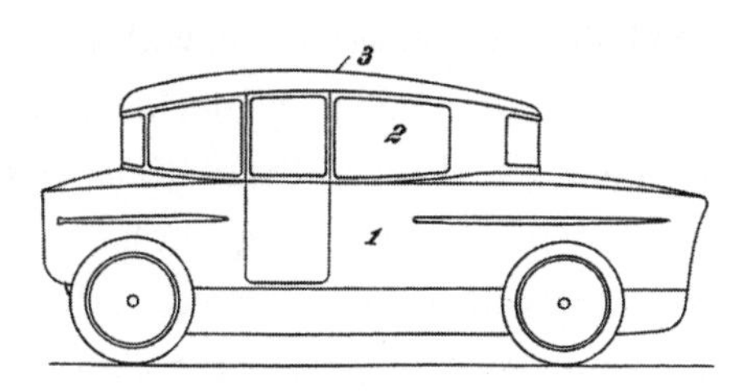

Patent drawing of the Rumpler Tropfenwagen, 1921

His plans faced immediate delay, however. In January 1922, Josef received from his mother in Vienna the shocking news that his father Hugo had died at the age of fifty-nine. Josef dropped everything, his studies and his innovative designs, and traveled to Vienna by train as quickly as he could to support his mother. She would die just four years later, aged only fifty-three. In February 1922, Josef had to undergo surgery in Frankfurt, as a long-term result of his war wounds. Once he had recovered, he returned to Vienna for several months.

Ganz-Klein-Wagen

Josef Ganz moved back to Darmstadt in the fall of 1922 to resume his studies and his car-building plans. He was increasingly interested in wave motion, whether of electromagnetic radio waves or the vibrations that occurred in mechanical suspension systems. He spent a considerable amount of time studying the way motorcycle and car wheels were sprung and concluded that, contrary to the general belief at the time, a car could hold the road perfectly well without being heavy. The drive wheels at the rear of an automobile, he decided, needed to be cushioned by the use of swing axles. This was the method used in the Rumpler drop car, with its two independently sprung rear half-axles, each of which could move in its entirety as the wheel bounced on the road. A car built in this way was immediately recognizable by the slightly raked angle of its rear wheels. With the correct design for the drive mechanism and suspension, he would be able to make his car as light as other factors allowed without any negative effect on road holding—a revolutionary approach to automobile design. It might well prove better and safer to drive than the large models produced by established manufacturers.

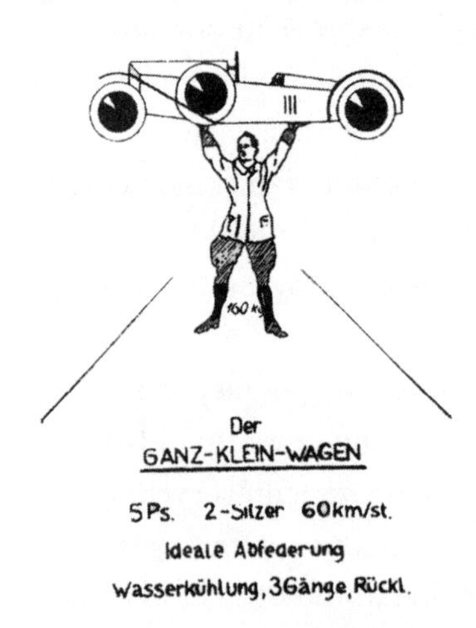

Prospectus for the Ganz-Klein-Wagen, 1923

In 1923, Ganz made his first drawings of a small four-wheeled car with a rear-mounted engine and all-round independent suspension. He even asked a Swiss architect friend to produce the drawings for a prospectus for what he called the Ganz-Klein-Wagen (a play on his own name, meaning "really small car").[15] On the cover of the brochure, he used a line drawing of a man holding a car barely bigger than his own body high above his head. Inside were two designs, one of which was for a streamlined, open-topped three-seater, with the driver sitting in the middle for optimum visibility and on either side of him a passenger seat set a little way back. The engine was behind the seats, just in front of the rear axle—as in the Rumpler Tropfenwagen—and all four wheels had independent suspension. The total weight of the car was calculated at a mere 160 kilograms (353 pounds), less than the combined weight of the three people in it.

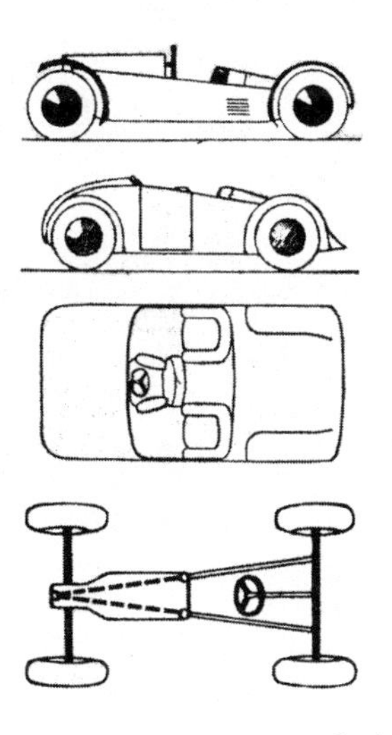

Prospectus for the Ganz-Klein-Wagen, 1923

Karl Herr, an old friend from Ganz's days in military service who worked at the Benz car plant in Mannheim, offered to help build a prototype.[16] Benz had just developed its own racing version of the drop car, and Herr shared Ganz's belief that there was a market for a small model along the same lines, especially now that Germany was having to contend with a serious economic depression and hyperinflation as a result of the occupation of the highly industrialized Ruhr area by French and Belgian troops, part of their effort to force Germany to make reparations payments.

In the fall of 1923, while Josef Ganz was working hard on the Ganz-Klein-Wagen and continuing his studies at the same time, he suffered exactly the kind of accident he had warned others about. On one of the unpaved roads outside Darmstadt, which he had ridden along dozens of times, the rear wheel of his motorcycle slid sideways in the mud. He lost control and fell badly, the heavy machine on top of him, and the engine block shattered his lower left leg. Moaning with pain, he lay in the muddy road under his bike until a passing motorist found him and took him to a hospital in the nearby city of Frankfurt. In a long, complex operation, the doctors managed to save the leg, but it would be a good two years before he recovered completely. Josef Ganz had become a victim of the motorcycle, precisely the danger he was hoping to prevent by creating a Volkswagen.

CHAPTER TWO

A NEW CALLING

Josef Ganz devoured stacks of books, newspapers, and magazines during the fall of 1923. The bones in his lower left leg had been crushed so badly in the motorcycling accident that the normally energetic Josef was bedridden for months after the operation, and his studies as well as his plans for a Volkswagen were put on hold. With his dog Wauzi to keep him company, he read the daily papers to keep up with the latest reports of growing political unrest in the country. The economic crisis was worsening. In September 1923, with hyperinflation running at a terrifying 2,500 percent per month, the government in Weimar announced a state of emergency. In the two months that followed, the basic cost of living rose so rapidly that a loaf of bread was soon priced at billions of marks, and it became cheaper to feed banknotes into the stove than to buy firewood. In his student room in Darmstadt, confined to bed, twenty-five-year-old Josef Ganz did his best to survive the insanity of it all. Just three years earlier, shortly before his death, his father Hugo had published a pamphlet in which he called on the creation of a world economical bond as "the last and only means to the salvation of civilization."

Deep in the south of Germany, in the midst of the chaos, the state of Bavaria was pressing for independence. Political tensions erupted in

Munich on the night of November 8, when SA stormtroopers commanded by Adolf Hitler, the thirty-four-year-old leader of the Nazi Party, mounted a coup aimed at toppling the socialistic Weimar Republic. The poorly organized coup attempt failed. Hitler was arrested and after a remarkably lenient trial was sentenced on April 1, 1924, by his right-leaning judges to five years in prison. He would serve only one, in unusually comfortable conditions in Landsberg jail. There he wrote his book *Mein Kampf*, which includes references to the ideas of the openly anti-Semitic Henry Ford, the man who had motorized America with his Model T Ford, introduced in 1908 and known as the "People's Car of the United States." Hitler was interested in Henry Ford mainly for his fiercely anti-Semitic beliefs and his rigid, almost dictatorial factory regime, while Josef Ganz believed Ford was demonstrating what Germany should do to develop a healthy car industry.

A number of European manufacturers tried to follow Ford's example by offering small, cheap models, but they could see no way to match the American firm's rate of production. In 1922, for example, Citroën in France launched its Type C 5 HP. It had a conventional design that was copied by Opel in Germany with its Opel 4 PS Laubfrosch ("Tree Frog"), nicknamed after its dark green color, the only distinct difference to the lemon-yellow-colored Citroëns. More revolutionary was the Tatra 11 of 1923, designed by the company's technical director Hans Ledwinka, with an air-cooled, two-cylinder boxer engine mounted at the front of a simple tubular-backbone chassis. The rear wheels had independent suspension and swing axles. Ledwinka's ideas were not entirely new. Like Edmund Rumpler's swinging half-axles, the backbone chassis had been patented more than twenty years earlier by the German inventor Dr. Georg Klingenberg,[1] although the concept had hardly ever been applied in practice. Josef Ganz had a great deal of respect for Ledwinka's achievement, and the Tatra 11, like the Rumpler Tropfenwagen, helped to inspire his own Volkswagen design.

The Golden Twenties

The French and Belgian occupying forces withdrew from the Ruhr area in 1924, and Germany entered a period of relative prosperity and political calm. After his long convalescence, Josef Ganz picked up his studies again, having missed three terms, and resumed work on his plans for a People's Car. He produced an improved design, using a backbone chassis largely inspired by the Tatra 11 but with the engine at the rear. His ghastly accident had reinforced his belief that the motorcycle, although cheap, was an inherently dangerous mode of transportation. He still had radical plans to build a safe, four-wheeled Volkswagen for the price of a good motorbike, but after long months in bed he was broke, unable even to pay for patent applications. He failed to hold on to the rights to an intake regulator for two-stroke engines, as described in his patent application of February 1923,[2] which was later used by many car manufacturers.

When he complained about this to his good friend Oskar Ursinus, the aviation pioneer and founder of *Klein-Motor-Sport*, to which Ganz continued to contribute occasional articles, Ursinus advised him to publicize his ideas.[3] If he lacked the financial means to patent his inventions, Ursinus argued, then he could earn money by writing about them. Ganz followed his friend's advice, and even before completing his studies he found himself writing more and more articles about innovative auto engineering. He described his Volkswagen concept in the magazine *Deutsche Motorsport Zeitung* in June 1924:

> I imagine mounting an ordinary 1/3 liter engine with a three-speed gearbox in the rear, which would drive one of the swinging rear half-axles. The simplest way of attaining the necessary rigidity of the vehicle as a whole would be by means of a long central tube, with the front suspension, pedals, seats, and drive mechanism attached to it.[4]

The idea was revolutionary in its simplicity, but, for the penniless student Josef Ganz, impossible to realize. He clung tenaciously to his concept, and in the 1924-5 academic year he specialized in car building. Although he had lost the rights laid down in his earlier patent application, he managed to secure patents on a board game with tiny cars that was used for advertising purposes[5] and on an electromagnetic gearbox.[6] His improving financial situation ran more or less in parallel with that of Germany, which like Ganz was slowly struggling to its feet after a major crisis. The economy picked up, and the political extremism of both left and right all but disappeared from the Reichstag, including the Nazi Party, whose leader Adolf Hitler had been released from jail in December 1924. It was the elites, however, that profited most from the economic boom of the "Golden Twenties." The rich lived decadent lives without a thought for the middle classes or the workers, who were by no means well off. The traditional class divide was reflected in the car industry, which to Ganz's dismay continued to concentrate on producing large, luxurious, conventionally designed cars for an elite clientele that could easily afford them, rather than looking for new ways to make cars lighter, safer, and cheaper. The introduction of a small Volkswagen seemed as remote as ever. The more affordable models on sale were mostly copies of much larger cars. Their driving qualities were unimpressive, and they were disproportionately expensive.

In the spring of 1926, in an effort to draw attention to the lack of technical innovation in German car building in general, Josef Ganz wrote a long article, asking "Has the motor vehicle industry reached its peak?" It was published in three installments in the auto magazine *Motor und Sport*. He discussed the serious design errors that were being made by manufacturers, whose basic method was to take an engine, build a chassis around it, and then adjust the bodywork to fit. Ganz argued that a car should instead be designed around the people inside it, as a carefully considered whole. He used diagrams to illustrate the best ergonomic position for the driver, to show where the center of gravity of a car should lie, and to indicate the main advantages of streamlining, lightweight alloys, and

Hanomag 2/10 PS Kommissbrot, 1925

independent suspension. "The greatest mistress of engineering is, after all, nature," he wrote. Man was merely "called upon to develop further what she has prepared and handed to us." In his article, Ganz went so far as to look at the psychological effects the shape of a car could have on a driver, with its "armoring" and the view over a long hood.[7] He wrote with huge respect about the ideas of engineers such as Edmund Rumpler, Paul Jaray (another pioneer of streamlining), and Hans Ledwinka of Tatra.

As well as the Rumpler Tropfenwagen and Ledwinka's Tatra 11, Ganz felt a great affinity for the Hanomag 2/10 PS Kommissbrot, introduced in 1925, a small car, designed by the German engineer Fidelis Böhler, with a rear-mounted engine, independent front suspension, and curved bodywork with integrated fenders and a smooth floor pan. Its unusual bodywork was the origin of its popular name, since it looked a little like a *Kommissbrot,* a cheap loaf of bread familiar to all Germans as a crucial component of army rations. All three of these cars, Ganz wrote, had revolutionary features that he wanted to incorporate into his Volkswagen.

Metropolis

With his three-part article in *Motor und Sport,* Josef Ganz made a reputation for himself as a critical and highly competent journalist. The Rumpler Tropfenwagen had been taken out of production a year earlier after extremely disappointing sales and risked going down in history as a failed experiment. After the fusion of Benz with Daimler in 1926, production of the Benz Tropfenwagen was suspended by its technical director, Ferdinand Porsche. The last two Rumpler drop cars ever built were bought by the German film director Fritz Lang for the role of "car of the future" in his science-fiction film *Metropolis,* which premiered in January 1927. Among those who squeezed into the cinema in Frankfurt one busy Saturday that month were Josef Ganz and Madeleine Paqué, his landlord's niece, who lived in Frankfurt along with her mother Maria Paqué-Köpp. Josef was thoroughly charmed by Madeleine, an elegant woman who dressed in the latest fashions, and invited her to go with him to the much-discussed film.

Ganz was fascinated by Lang's vision of an industrial city with elevated highways and planes flying between immense skyscrapers, but he did not see the Tropfenwagen as a futuristic design for 2026, the year in which the film was set. As he had written in *Motor und Sport*, to him it was a perfectly realistic concept that set an example to the industry and ought to be rolling off the production line in huge numbers in a range of different models and sizes. So it came as a shock to see the final scene of the two-and-a-half-hour masterpiece, in which the two Tropfenwagens went up in flames on the funeral pyre of the "goddess of doom," the robot-woman Maria. As Josef Ganz and Madeleine Paqué sat talking afterwards in a nearby restaurant, he explained to her why the ending seemed to symbolize the spirit of the times: innovation was not encouraged; in fact, it was thrown onto a pyre. He was determined to reform the auto industry. Although she found him a little eccentric, Madeleine was fascinated by the passionate Josef. They talked for hours and late that evening he dropped her back at her mother's stately villa. It was the beginning of a decades-long romance.

Secondhand Hanomag

On March 4, 1927, the moment arrived that Josef Ganz had been working toward for so many years. After all the delays caused to his studies by the war, his motorcycle accident of 1923, and the deaths of his father in 1922 and his mother in 1926, he was finally awarded his official diploma in mechanical engineering by the Technical College in Darmstadt (today's Technische Universität Darmstadt). He was now Dipl.-Ing. Josef Ganz. Despite his many setbacks, Ganz achieved the best marks of any student since the college was founded in 1877.[8] The town's adult education center offered him a job as a teacher of auto and radio technology, and he was also taken on as a German sales representative by the radio manufacturer Ingelen in Vienna, for which he developed new radio sets.[9] Now that he was finally earning a proper salary, he wanted to buy a car. He came upon a secondhand Hanomag 2/10 PS Kommissbrot at a dealership that spring and bought the little white convertible with wooden-spoked wheels. Bruised and battered from his many falls, he took his leave of motorcycling for good. That summer, Josef Ganz strapped his leather suitcases to the back of the little rear-engined Hanomag, said goodbye to his beloved Madeleine, and left for Switzerland to take an extensive tour of the Alps.

Josef Ganz had been visiting Switzerland regularly since childhood, staying with his uncle Alfred Ganz in Luzern, and he knew the dangerous Alpine passes all too well. That summer he must have been planning to put his Hanomag through the ultimate test, to discover all the strengths and weaknesses of its design. Exhaustive test drives were not yet standard practice. It was left to the buyer to discover a car's faults and limitations. Many models revealed their weak points only after they had been driven extensively. Ganz encountered the first problems with his Hanomag when he tried to start it up. It had no electric starter motor, which meant having to crank it every time, and as soon as the engine ignited, the crank shot back violently, almost skinning his fingers. The only way to reduce this effect was to adjust the ignition almost weekly, which brought him to a second practical disadvantage: the engine space was difficult to access through the small hood; some parts of the mechanism were extremely awkwardly sited. Josef Ganz decided it would be far more practical if the entire back shell of the vehicle hinged up, making the engine accessible from all sides. He sometimes spent hours at the side of the road removing parts that were in the way before he could carry out repairs. Aside from these irritations, he was pleasantly surprised at how well the Hanomag behaved on the hairpin bends of mountain passes like the Oberalp, Julier, and Bernina. On the verges and in the ravines, innumerable wrecks lay as silent witnesses to the fates of others on those treacherous roads.

The weaknesses of his Hanomag were not, Ganz concluded, the result of poor design in general but of the fact that the car was "not yet fully thought through in detail."[10] At the top of the Oberalp Pass, he met up with his sister Margit, who had traveled to Switzerland from Vienna with her husband Jakob Feldhammer, and spent several days with the couple, but after a wonderful trip lasting several weeks, Josef was forced to return to Germany that August. He had run out of money and needed to prepare for his new teaching job at the adult education center in Darmstadt, where the new term would shortly begin. After all his adventures with the Hanomag, Ganz decided to trade it in for a slightly larger model, something more comfortable and reliable. He knew of only one good option: the Tatra 11. During his holiday, he had noticed how well the Tatra, with its air-cooled engine, performed in the mountains, passing many other vehicles stranded at the roadside with boiling radiators. In late summer that year, Josef Ganz proudly bought a brand-new Tatra 11, on credit, from the car dealer Donges und Wiest in Darmstadt.

In the evenings, after classes were over, Josef Ganz continued further perfecting his Volkswagen concept, informed by his experiences driving the Hanomag and the Tatra. He often withdrew to his room after dinner to work on the design, using enormous sheets of stiff white paper at his drawing table, lit by a single bulb. He made 1/5 scale drawings of the entire vehicle and detailed diagrams of elements such as the independent suspension with swinging rear half-axles and the air-cooled, two-cylinder boxer engine, built as a single unit with the gearbox.[11]

Klein-Motor-Sport

In December 1927, Josef Ganz's career suddenly changed direction when he received a telephone call from the secretary at H. Bechhold Verlagsbuchhandlung, the publisher of the magazine *Klein-Motor-Sport*. She asked whether Ganz would be able to come to Frankfurt very shortly to talk with the publishers, Wilhelm Breidenstein and Dr. Hartwig Breidenstein. He agreed immediately and a few days later made the trip from Darmstadt to Frankfurt in his Tatra. He got along well with the congenial Hartwig and his elder, rather more formal brother Wilhelm, who told him they were looking for a new editor-in-chief, someone with the enthusiasm required to breathe fresh life into *Klein-Motor-Sport*. The magazine had been published irregularly of late, by a different publisher, and the number of subscribers had fallen to a mere three hundred. The Breidensteins had heard about Josef Ganz from Oskar Ursinus and were impressed by his passion in confronting the German auto industry in critical articles over the past five years. The three men discussed the lack of technical innovation in the industry, and Ganz described his own way of thinking and his vision of a People's Car. He said that one major cause of the decline of "individual German car building" was undoubtedly the enormous power that lay with the financiers, while the task of an engineer was confined to implementation.[12] The Breidenstein brothers were so impressed that they immediately appointed Ganz editor-in-chief, allowing him considerable freedom to decide on the content of the magazine. Beginning in January 1928, he was able to use the periodical as a platform from which to bombard the industry with criticism and at the same time describe his own innovative concepts. Ganz drove back to his house in Darmstadt bursting with enthusiasm, decided to give up his job at the adult education center right away, and set about developing ideas for *Klein-Motor-Sport.*

During the Christmas and New Year holidays and on into 1928, Ganz worked hard at producing copy for the first issue to be published under his editorship, which was due to appear in mid-January. He was based in a soberly furnished office in his house in Darmstadt, with little more than a wooden desk and chair, a filing cabinet, a radio with a large round Bakelite loudspeaker, and a telephone. He had a new connection installed, and the phone number for the editorial office of *Klein-Motor-Sport* became Darmstadt 544. Keen on photography and technical drawing, Ganz wanted to create most of the illustrations for the magazine himself. He produced the black-and-white line drawings on his drawing table, while the bathroom served as a darkroom for the developing and printing of photographs. In the first few weeks of 1928, Madeleine Paqué helped him to get all the articles and illustrations ready for the printer, using a redesigned layout. This was an extremely exciting and challenging time for Josef Ganz. From the start, he had clear aims in view for *Klein-Motor-Sport.* It was to become a platform open to anyone who wanted to promote technically progressive ideas or to express well-founded criticism of poorly designed cars. This was one way to promote the development of a technically advanced, cheap German Volkswagen, and Ganz also planned to use the magazine as a weapon against corruption in the industry and the media. An advocate of uncensored freedom of expression and complete openness and transparency in the industry, he was disgusted by the way the majority of auto journalists practically ate out of the hands of established manufacturers, delivering fulsome praise for hopelessly outdated models.

For the German People's Car

The first issue under his leadership opened with an analytical article by Josef Ganz on the state of the economy in post-war Germany, where car ownership was still the height of luxury, entitled "*Für den deutschen Volkswagen*" ("For the German People's Car"). How different things were in America, he wrote. There Henry Ford had introduced his immensely popular Model T in 1908, putting it into mass production two years later, while in Germany and many other European countries, cars were still constructed according to traditional methods, by hand. This made them expensive both to buy and to maintain, and the automobile had yet to penetrate to the lower classes of the population. What was lacking in Germany was "the German or in a broader sense European Volkswagen." "It's not at all utopian, as the many failures in this field might at first make us think. The problem is solvable to everyone's complete satisfaction, if tackled in the right way."[13] Ganz had firm opinions about what was needed. His opening salvo was followed by the first installment of a three-part article in which he described what the future Volkswagen should look like. The final part included a picture of what he conceived as the cheapest version of the Volkswagen: an open-topped, streamlined two-seater. The car had a low chassis with all-round independent suspension, and the engine and gearbox were just in front of the rear axle, which meant that the center of gravity was low and in the middle of the car. Apart from a good technical design, mass distribution of the Volkswagen would require the law to be changed to allow people to drive it with a motorcycle license, since hardly anyone was in possession of a driver's license for a car. New legislation to that effect came into force in April 1928, but it applied only to three-wheeled cars weighing no more than 350 kilograms (around 772 pounds) and with a maximum engine capacity of 0.2 liters, not to the much safer four-wheeler. Josef responded: "They don't even lift a little finger to help us."[14]

For the remainder of 1928, Josef Ganz threw himself heart and soul into *Klein-Motor-Sport*. He visited auto exhibitions, took countless test drives, and established contacts with auto designers and manufacturers at home and abroad. Instead of writing exclusively about recently introduced models, he included in his articles numerous suggestions for future improvements, repeatedly stressing the importance of technical innovations such as independent suspension, swing axles, the backbone chassis, the positioning of the engine adjacent to the driven front or rear wheels, a low center of gravity, and streamlining. He published technical drawings of axle and chassis designs, for example, or of the shape of the "car of the future": low, teardrop-style, streamlined, with the engine mounted in front of the rear axle.[15] Only later would Ganz realize that in his articles he had sometimes given away valuable, patentable ideas.[16] One new model that did meet with his approval was produced by the Röhr-Werke in Ober-Ramstadt, a creation of its director Hans Gustav Röhr and his technical team, led by the engineer Joseph Dauben. Ganz was deeply impressed after taking it for an extensive test drive in the spring of 1928 and barely knew how to convey his experience to his readers. "For anyone who hasn't driven the car himself it will be impossible to imagine its holding on poor roads, in bends, and when braking sharply, and its power of acceleration," he wrote in *Klein-Motor-Sport*. He believed the Röhr car's unique handling was entirely attributable to its "revolutionary design," which incorporated all-round independent suspension with swing axles at the rear, front-wheel-drive, lightweight all-metal bodywork, and a low center of gravity. Josef concluded that it was "the only German state-of-the-art car."[17]

To Switzerland in a Dixi

That summer, Josef Ganz decided to return to the unforgiving Swiss mountain passes to which he had subjected his old Hanomag the year before. The directors of the Fahrzeugfabrik Eisenach had agreed to his request to be allowed to make the same demanding journey in a Dixi 3/15, a version of the British Austin 7 that they were building under license. The car provided by the factory had already clocked up a considerable distance. Even before leaving Eisenach, Ganz noticed that the brakes were barely functioning and the steering was becoming heavier by the mile. The 125-mile trip back to Darmstadt on poor roads was a wild ride. The car drifted all over the place. It was unpleasantly reminiscent of his hazardous motorcycling days. At home in Darmstadt, he carried out repairs to the little Dixi and over the next few days drove well over six hundred miles in the nearby countryside before leaving for Switzerland via Freiburg in southern Germany in August.

In Freiburg, Ganz and his Dixi took part in a mountain race called the Freiburger Bergrekord-Rennen, and he met up with his good friend Frank A. E. Martin, editor-in-chief of *Motor und Sport*. It was five in the afternoon by the time he left Freiburg to drive non-stop to his Uncle Alfred's villa in St. Niklausen near Luzern. Along the way, countless motorcyclists passed him, returning from the race. The light was fading as he drove away from the border post near Basel, and slowly absolute darkness came down, so that he had to rely on the little Dixi's small headlamps. At least they were a good deal more effective than the single headlamp on his old Hanomag, which had left him virtually feeling his way round bends. To add to the demands on his concentration made by the blackness on all sides, the heavens opened with an almost tropical downpour. At around eleven thirty that evening, he arrived at the village of St. Niklausen, completely drenched in the open-topped car, and turned into the driveway of the Villa Solina. His uncle and aunt, Alfred and Valerie Ganz, who had watched him approach through the darkness, ran out to greet him. After several enjoyable hours with his uncle, Josef flopped into bed dead tired. Next morning, he showed his uncle and aunt the little Dixi he had driven all the way from Darmstadt to Luzern. Amazed by the tiny car, Alfred Ganz joked that Josef's Tatra had undergone quite a metamorphosis, and Valerie tried to park it in the large dog kennel at the bottom of the garden.

Several days later, Ganz left for the Alps to take the ultimate test drive. In the Rhône valley, he came upon a curious spectacle. A driver had changed down too late on the treacherous mountain roads in his Panhard and ended up on the wrong side of the wooden crash barrier. Josef felt sorry for the poor man, who because of the heat had been at the wheel in only his shirt, underpants, socks, and garters. There he stood, half naked at the roadside, taking the flak from his furious wife and three daughters. Ganz expressed solidarity with the unfortunate fellow and helped him until eventually a horse was brought in to drag the heavy Panhard up the mountain to the nearest village, one tortuous step at a time. Accidents were commonplace in the busy summer months, and the sides of the roads were scattered with the wrecks of cars that were no match for the mountains—grim evidence of the skill a motorist needed to keep a technically inferior car on the road in such precipitous terrain. The exhaustive test drive made by Josef Ganz in his little Dixi was so unusual in those years that he decided it was worth dedicating a whole issue of *Klein-Motor-Sport* to it. He was full of praise for the car and advised the manufacturer to draw on the knowledge of "experienced designers" such as Hans Ledwinka or Gustav Röhr if it wanted to avoid "losing its lead."[18]

Friends of Göring

Back in Darmstadt, Ganz took his Tatra out of the garage again, but as he was driving a technical problem emerged. A ball bearing in one of the swinging rear half-axles kept jamming. He took his guarantee to the car dealership Donges und Wiest, where he had bought it, but because of the complexity of the problem he was referred to DETRA (Deutsche Licenz Tatra-Automobile Betriebsgesellschaft mbH) in Frankfurt, which assembled the car under license. There he met Paul Georg Ehrhardt, a technician who shared his ambition to build according to non-standard methods, using a lightweight structure, swing axles, and streamlining. Ehrhardt was close to the auto manufacturer Gustav Röhr, who, like him, had been a fighter pilot during the war alongside the air ace Hermann Göring, a good friend of Ehrhardt's who was elected to the Reichstag in May 1928 as one of twelve Nazi Party members of parliament. Ehrhardt had also worked with Albert Einstein on wing designs and had several patents for car and aircraft technology to his name. After the 1914-18 war, Ehrhardt had been employed by Luftschiffbau Zeppelin in Friedrichshaven at the time when Paul Jaray was carrying out his first experiments with streamlined automobiles in the wind tunnel there. Ganz was impressed by Ehrhardt's experience and progressive ideas and suggested he should become an editor at *Klein-Motor-Sport*.

Despite Ehrhardt's help, DETRA could not fix the Tatra 11, and Josef was referred on to the Tatra factory in Czechoslovakia.[19] In September 1928, he set out on the long journey from Darmstadt to Kopřivnice, where he was politely received by the directors of Tatra and given a room at the factory hotel while the problem with his car was investigated. He took the opportunity to meet and exchange ideas with the technical director, Hans Ledwinka, a man he called "the supreme master of European car building."[20] Tatra was one of a handful of auto manufacturers in Europe that had more or less turned their backs on the standard methods of making cars, the others being Austro-Daimler and Steyr in Austria and Röhr in Germany. Josef Ganz was so enthusiastic about the models being produced at the factory that on his return from Czechoslovakia he devoted almost an entire issue of *Klein-Motor-Sport* to the need for technological innovation. In it he made an appeal: "No more cars without swing axles!" Compared to "live" or solid axles, they offered so many advantages that Ganz was convinced that "very shortly any vehicle without swing axles will be considered ready for the scrapheap."[21] He also emphasized once again the great value of the Rumpler Tropfenwagen concept. Despite the fact that it now seemed more or less forgotten, it was "absolutely healthy, just as healthy as Böhler's Hanomag, which has likewise been condemned to death."[22] The death sentence for the Hanomag 2/10 PS Kommissbrot was carried out at the end of 1928 with the introduction of the new Hanomag 3/16 PS, a thoroughly conventional model with a front-mounted, water-cooled engine and a long drive shaft to the rear wheels. Josef Ganz was aghast that after producing the Kommissbrot for four years Hanomag was now reverting to the standard design that he was so passionately trying to eradicate with his criticism in *Klein-Motor-Sport*. Hanomag claimed that the shape of the Kommissbrot did not appeal to potential buyers. The public, they said, wanted "a real car."[23]

The "swing-axle issue" of *Klein-Motor-Sport* was a great success, and the magazine was now rapidly gaining influence. It fell to the postman in a quiet residential area of Darmstadt to deliver sackloads of supportive letters to the editorial office. Clearly filling a gap left by the other specialist journals, Ganz was rewarded with a growing pool of readers, although at the cost of increasing opposition from conservative industrialists, who would have liked to put the magazine out of business. An anonymous letter writer in Frankfurt scornfully described Ganz as a "Daitscher," German-Yiddish for an immigrant Jew, adding "perhaps from Galicia,"[24] a reference to the former principality in Central Europe that was home to a large Jewish community. It was a sign of the alarming anti-Semitism that had been simmering below the surface in Germany for many years and was now being capitalized upon gleefully by Adolf Hitler and his Nazi Party.

CHAPTER THREE

MOTOR-KRITIK

In mid-December 1928, Darmstadt and the surrounding countryside looked like something out of a fairy tale. The streets, with their majestic houses, tall trees, and the occasional parked car, were carpeted with a deep layer of virgin snow. Josef Ganz warmed himself thoroughly by the stove with a cup of strong coffee before heading for Frankfurt. It was a drive of only twenty-five miles, but the traffic was moving at little more than walking pace on the slippery streets. He went out to his car wrapped in a thick leather jacket, long scarf, and sturdy gloves. After cranking the air-cooled engine of his Tatra, he hurriedly got in, tramped the gas pedal, and set off. During the slow, icy drive, continually having to wipe condensation from the windows, Ganz thought back over the turbulent developments of the past year. The publishers of *Klein-Motor-Sport* had invited him to Frankfurt to discuss his work as editor-in-chief and how he intended to pursue it in the course of 1929. He was confidently looking forward to the meeting, since both the reputation of the magazine and the number of subscribers was growing by the day. To illustrate the fact, he had brought with him a letter from a reader in Berlin who had direct experience of the success of the publication. Half an hour after the latest issue reached the shops, it had sold out at his local newsstand, he wrote, and within two hours it was unavailable practically anywhere in the city. The letter writer called on Ganz to increase the print-run a hundredfold.[1]

The Breidenstein brothers were deeply impressed by his progress, his competence, and his enthusiasm; in fact, they believed that under Ganz's leadership *Klein-Motor-Sport* could become an even bigger success. They offered him a four-year contract, to run from January 1, 1929, through December 31, 1933, with a monthly salary of seven hundred Reichsmark. The contract explicitly stated that *Klein-Motor-Sport* was not obliged to follow the general motoring news the way other specialist publications did but should distinguish itself by "exercising constructive criticism of the car and motorcycle industry as a whole," while at the same time providing "factual, lively articles, sometimes with illustrations, to inform the technically interested layperson."[2] It was then that Ganz proposed giving a more powerful aura to the unique direction the magazine was taking by changing its name to the more appropriate *Motor-Kritik*. The Breidenstein brothers eagerly agreed; in fact, they were so convinced of the glowing future that awaited the magazine under its new title that they decided to raise the price of single issues from 0.50 to 0.60 Reichsmark and the price of a quarterly subscription from three to 3.60 Reichsmark.[3] Josef Ganz now had the financial means at his disposal to employ his girlfriend Madeleine Paqué on a permanent basis as his editorial secretary, to take on additional freelance journalists, and to include far more photographs and drawings than had ever appeared in the rather soberly produced *Klein-Motor-Sport*.

Car of the Future

Josef Ganz left the Breidensteins' office in high spirits and drove back through the cold to Darmstadt to focus all his attention on *Motor-Kritik*. The articles he wrote for the January 1929 issue included a description of the basic design of the future Volkswagen, a piece about an innovative prototype of a continuously variable gearbox, and an article examining a revolutionary prototype of a car designed by the Czechoslovak engineer Willibald Gatter. Before setting up in business for himself, Gatter had worked at Austro-Daimler for several years under its technical director, Ferdinand Porsche. In 1922, he had assisted Porsche in the development of a lightweight racing car named Sascha, after the Austrian film pioneer Count Alexander Joseph "Sascha" Kolowrat-Krakowsky, a good friend of Porsche and a man with a great passion for cars. Count Kolowrat, a keen racing driver, had built his own cyclecar in 1913 and in the early 1920s had tried to persuade Porsche at Austro-Daimler to put a "Volksauto" into production. Three prototype open-topped cars had resulted, built to a conventional design, as well as the Sascha racing model, but none went into serial production. Willibald Gatter continued to believe in the idea of a People's Car, and in 1926 he came up with a simple open-topped version that he intended to put on the market at a low price. He also developed a larger model, and it was this that Josef Ganz decided to write about in *Motor-Kritik*.

Ganz had arranged by telephone that he and Gatter would meet in Darmstadt in early January 1929 to test the new prototype. They got along well, and with Madeleine Paqué and her mother in the back seat they took the large Gatter convertible for a thorough test drive, during which Gatter demonstrated all its special features. Like Ganz's design for the Volkswagen, the car had a backbone chassis, independent suspension, and swinging rear half-axles. Its front-mounted engine drove the rear wheels. In his article for *Motor-Kritik*, Ganz wrote that "in a time so lacking in purposeful automobile designers," it was a model for the "European car of the future."[4]

By mid-January, the new issue was ready. Across Germany and beyond, the much-discussed *Klein-Motor-Sport* dropped onto the doormats of hundreds of car and motorcycle manufacturers, engineers, technicians, and other subscribers. Beneath the old title, in a powerful, specially designed font, in large bright-red letters that took up the full width of the cover stood the new name: *MOTOR-KRITIK*. It would keep the double title for a transition phase of one year. The magazine radiated enthusiasm for the future and promised the reader "important technical advances in motor vehicle construction." On the opening page, the publishers Hartwig and Wilhelm Breidenstein expressed their support for "the crucially necessary free exchange of views in criticism" to which the periodical aspired:

> A year ago, *Klein-Motor-Sport* threw off the manacles so willingly worn by most of the trade press, and as an outward sign of this it has adopted the new name *Motor-Kritik*. Criticism alone is sterile, unless at the same time better ways are indicated. We therefore regard that as our primary task. Fellow workers, friends! Let us all cooperate to bring rapid recovery to the motor vehicle industry![5]

The "healing of the industry," as the publishers called it, was sorely needed. Practically everyone agreed that German car making was in a deep state of crisis, but opinions differed as to the cause. According to the car industry's trade association, the Reichsverband der Automobilindustrie (RDA), the blame was entirely attributable to foreign imports and their detrimental effect on German production, but Josef Ganz and his colleagues at *Motor-Kritik* believed the cause lay in the fact that German manufacturers had stuck to conventional, sometimes downright poor designs and outdated production techniques, losing customers as a result. They believed the RDA itself was at fault for perpetuating the situation. Unlike its director, Privy Councilor Robert Allmers, Josef Ganz was not so nationalistic that he wanted to protect German manufacturers at all costs; in fact, he publicly attacked conventional and old-fashioned German car companies and was happy to publish advertisements by foreign manufacturers. He was widely criticized for doing so, but, as he saw it, his only mission was to stimulate technical innovation right across Europe and to encourage new ways of thinking, not necessarily to defend the interests of German car plants. Instead of encouraging innovation, the RDA was desperately trying to protect the German auto industry from negative re-

MOTOR-KRITIK

New name and logo: *Motor-Kritik*

porting. In early November 1929, it sent a letter to all the manufacturers under its wing with a contract for them to sign that obliged them to place no more advertisements with certain blacklisted periodicals, of which a list was enclosed. They included *Motor-Kritik*.[6] Josef Ganz published an irritated reaction to this advertising boycott by the "dictators" at RDA in his magazine: "What the Reichsverband der Automobilindustrie has built up here is an organization for the total blockade of any magazine whose existence does not suit a member of the Manufacturers' Commission."[7] The RDA was trying to undermine the financial position and therefore the influence of publications like *Motor-Kritik*. In a later issue, Josef Ganz protested once more by adding a yellow paper to the cover of *Motor-Kritik* with red text: "As a magazine true to its faith boycotted by the Reichsverband der Automobilindustrie!"

Als überzeugungstreues Blatt

vom Reichsverband der Automobilindustrie

Message added to the cover of *Motor-Kritik* 16, 1930

Modernizing Generation

Josef Ganz and Madeleine Paqué came into contact with a generation of young engineers who had found in *Motor-Kritik* an outlet for their progressive ideas. They developed friendships with many of them. As well as Willibald Gatter, they got to know the streamlining pioneer Paul Jaray, who had been living in Switzerland for the past five years. Jaray had patented his designs in 1921, and several prototypes were built in 1922 and 1923, but no manufacturer was willing to put a revolutionary streamlined model into serial production. Josef Ganz countered this conservative mentality by regularly emphasizing in *Motor-Kritik* the importance of the streamlining principles described by Paul Jaray and Edmund Rumpler. A quite different kind of innovation was suggested by Armin Drechsel, a German engineer and friend of Ganz who wanted to achieve improved road holding by stabilizing cars using a gyroscope.[8] He had patented his system and was working on a scale model to test how the "gyroscope car" would work. Another young engineer who became a regular contributor to *Motor-Kritik* was the Hungarian Béla Barényí. Like Josef Ganz, he had designed a "Volkswagen of the future" in the mid-1920s while completing his studies, with a backbone chassis, a rear-mounted boxer engine, and independent suspension, but had not succeeded in producing a prototype. In *Motor-Kritik,* Barényí wrote: "This way of building only according to the latest ideas should not be applied to isolated types of vehicle but generally recognized as the right way!"[9] The manufacturers were far from ready for such passionate resolve. They had designed and built cars according to their own familiar, standard methods for many years, and now suddenly the new magazine *Motor-Kritik* had emerged, led by an ambitious editor-in-chief called Josef Ganz, aged thirty-one, who was going to tell them how to do it all differently.

Logo of the streamline car pioneer Paul Jaray

Sabotage of the German Economy

The often cynical and harshly critical tone in which journalists writing for *Motor-Kritik* picked apart what they saw as hopelessly outdated models created ill feeling in the auto industry and led to a series of clashes with manufacturers. Established firms tried desperately to prove that the concepts propagated in the magazine were technically unworkable and did all they could to depict the magazine as an unprofessional, sensationalist rag. The German auto manufacturer Stoewer was one of the first to take this line, publishing a technically inept article[10] in its company

Josef Ganz at his drawing board, working on the swinging-axle design of his first Volkswagen prototype for the Ardie motorcycle company, Darmstadt, 1930.

Father of Josef Ganz, Hugo Markus Ganz (right), and his brother Alfred, around 1915.

Above: Margit, the sister of Josef Ganz, Vienna, 1907.
Right: Bust of Margit Ganz made by Gustinus Ambrosi in 1913.

Josef Ganz on his motorcycle in the early twenties.

Above: Josef Ganz as a teenager with his first patent, featured in the magazine *Stadt Gottes*, 1910.

Right: Josef Ganz's driving license, issued in 1920.

Left: Madeleine Paqué, the girlfriend of Josef Ganz, around 1927.
Above: Josef Ganz and Madeleine Paqué with their German shepherd Dolly, Frankfurt, 1931. Madeleine is behind the wheel of Ganz's second Volkswagen prototype, the May Bug.

Dolly visits Josef Ganz on his sickbed after a car crash in 1930.

Top left: Josef Ganz with his dog Wauzi in Darmstadt, 1925.

Top right: Dr. Michael Max Munk, who worked with Josef Ganz in Germany during the First World War, standing in his United States NACA office, 1926.

Bottom: Edmund Rumpler, designer of the Tropfenwagen, reading a copy of Josef Ganz's *Motor-Kritik*, with Madeleine Paqué and her mother Maria, all in the back seat of Josef's Tatra.

Streamlining pioneer Paul Jaray joins Josef Ganz (behind the wheel) in his second Volkswagen prototype, the May Bug, Frankfurt, 1931.

A streamlined 1928 Chrysler prototype by Paul Jaray, shortly after a collision with a cyclist, showing the safety aspects of the streamlined shape, 1933.

Josef Ganz's friend Armin Drechsel with a scale model of his "gyroscope car," around 1930.

Ferdinand Porsche in 1923.

Hungarian engineer Béla Barényí, who, like Josef Ganz, designed a Volkswagen.

Top left: Hans Nibel, technical director of Daimler-Benz, during a car rally on the Nürburgring, around 1930.

Top right: Wilhelm Kissel, director of Daimler-Benz.

Bottom: A 1923 Benz 11/40 Runabout with a broken front axle, attibutable, according to Josef Ganz, to the heavy and old-fashioned design.

Top: The 1923 Benz Tropfenwagen racing car was based on the revolutionary Rumpler Tropfenwagen.

Middle: The streamlined teardrop shape of the 1920s Benz Tropfenwagen.

Bottom: The chassis of the 1923 Benz Tropfenwagen, showing the mid-mounted engine and lightweight chassis construction.

The Swiss Alps, 1927: Josef Ganz test driving his Hanomag 2/10 PS Kommissbrot. His sister Margit and her husband Jakob Feldhammer join him in their own car.

Top: Josef Ganz prepares a Dixi 3/15 for a test drive through the Alps. Parked in front is his Tatra 11, 1928.
Middle: The backbone chassis of the 1923 Tatra 11 with swinging rear half-axles, designed by Hans Ledwinka.
Bottom: Hans Ledwinka, technical director of Tatra, during a company visit by Josef Ganz, 1928.

Top: German engineer Hans Gustav Röhr with his progressive front-wheel-drive car, around 1928. A design admired by Josef Ganz.
Bottom: Czechoslovakian engineer Willibald Gatter in his prototype during a visit to Josef Ganz. Madeleine Paqué and her mother Maria sit in the backseat, 1929.

Josef Ganz takes his first Volkswagen prototype for the Ardie motorcycle company for a test drive through the woods, 1930.

Top and next page top: The backbone chassis with mid-mounted engine and independent suspension of Josef Ganz's first Volkswagen prototype in the factory of the Ardie motorcycle company.
Bottom: The 1930 Ardie-Ganz prototype was given a simple "bug-shaped" open metal body.

Middle left: The rear of the Ardie-Ganz prototype with horizontal air intake, 1930.

Middle right: Josef Ganz takes the chassis of the Ardie-Ganz prototype for a test drive.

Bottom: The backbone chassis of the Ardie-Ganz prototype at the factory, 1930.

Top: Willy Bendit, director of the Ardie motorcycle company, along with a mechanic inspects the second more conventional body design of the 1930 Ardie-Ganz prototype.

Bottom: The second body design of the Ardie-Ganz prototype featured a fake, conventional-looking radiator to please prospective buyers.

Above: The sporty "boat tail" of the second Ardie-Ganz prototype body design.

Josef Ganz's Tatra parked behind the moving vans from Darmstadt in front of his new home and office in Frankfurt, 1931.

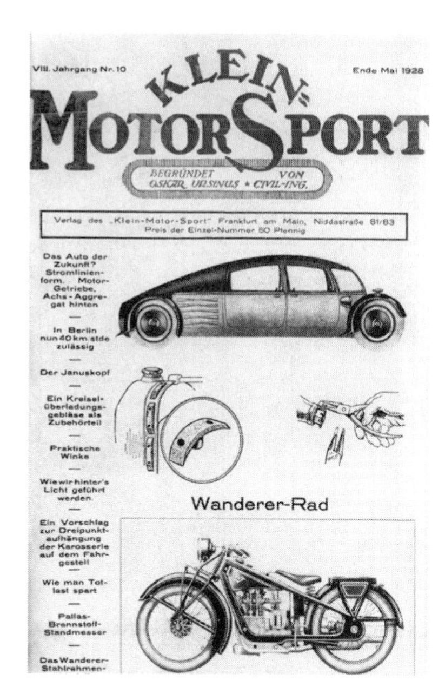

VIII. Jahrgang Nr. 10

Ende Mai 1928

KLEIN-MOTOR SPORT

BEGRÜNDET VON OSKAR URSINUS • CIVIL-ING.

Verlag des „Klein-Motor-Sport" Frankfurt am Main, Niddastraße 81/83
Preis der Einzel-Nummer 50 Pfennig

Das Auto der Zukunft? Stromlinienform. Motor-Getriebe, Achs-Aggregat hinten

—

In Berlin nun 40 km stde zulässig

—

Der Januskopf

—

Ein Kreisel-Überladungsgebläse als Zubehörteil

—

Praktische Winke

—

Wie wir hinter's Licht geführt werden.

—

Ein Vorschlag zur Dreipunktaufhängung der Karosserie auf dem Fahrgestell

—

Wie man Totlast spart

—

Pallas-Brennstoff-Standmesser

—

Das Wanderer-Stahlrahmen-

Wanderer-Rad

Klein-Motor-Sport with streamlined "car of the future" design by Josef Ganz, 1928.

JAHRGANG NR. 20

ENDE OKTOBER 1929

KLEIN-MOTOR SPORT

MOTOR-KRITIK

VERLAG FRANKFURT A. M., BLÜCHERSTRASSE 20/22 EINZELHEFT RM —.60

Starre Achs? Urzeit-Dachs!

Motor-Kritik cover on the advantages of independent wheel suspension, 1929.

Josef Ganz at his desk in the editorial office of *Motor-Kritik*, around 1933. He wears a wristwatch with a wire-frame cover to protect it during test runs.

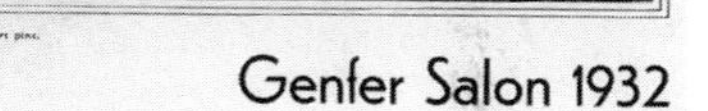

Top left: Motor-Kritik with rear-engined car by the Martin Aircraft Comp. in New York, 1932.

Top right: Motor-Kritik showing a Röhr to promote the Geneva motor show, 1932.

Bottom left: Cover of *Motor-Kritik* promoting streamlined car design, 1932.

Bottom right: Motor-Kritik with a streamlined car prototype designed by Paul Jaray, 1932.

Top left: Madeleine Paqué, girlfriend of Josef Ganz and secretary of *Motor-Kritik*, in the *Motor-Kritik* editorial office, around 1933.
Top right: Cover of *Motor-Kritik* dedicated to "roads and the people's car," 1933. This was the last issue of *Motor-Kritik* before Josef Ganz was arrested by the Gestapo.
Bottom: Delivery van from *Motor-Kritik* publisher H. Bechhold Verlagsbuchhandlung in front of the home and office of Josef Ganz, around 1932.

Top: Georg Ising, editor of *Motor-Kritik* and friend of Josef Ganz, before joining the magazine, 1920s. Ising became editor-in-chief of *Motor-Kritik* after Josef Ganz was arrested by the Gestapo in 1933.
Bottom left: Motor-Kritik showing the Auto-Union racing car designed by Porsche, published when Ganz was trapped in Switzerland, having narrowly escaped assassination, 1934.
Bottom right: Motor-Kritik showing two streamlined car prototypes by Paul Jaray, 1935.

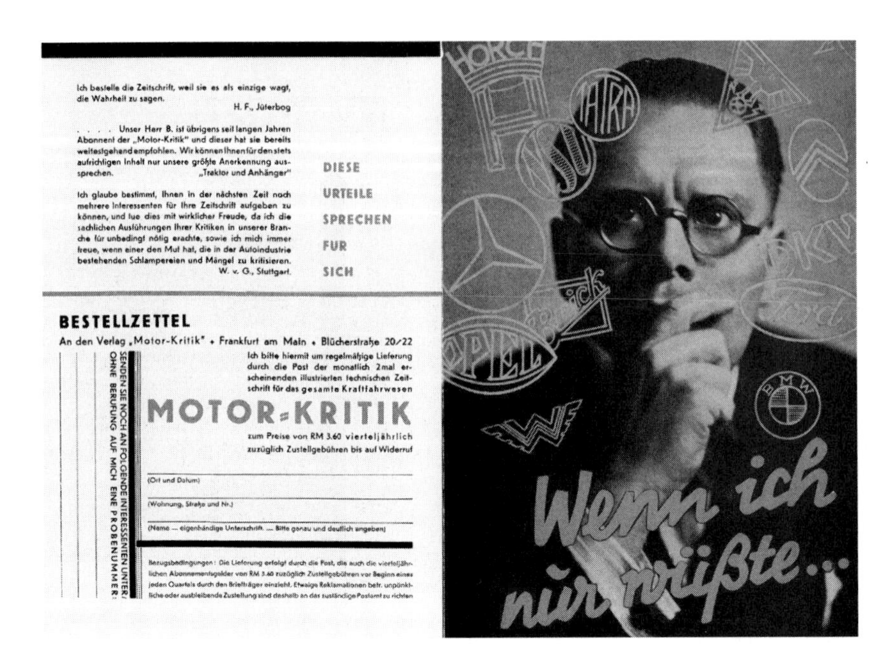

Ich bestelle die Zeitschrift, weil sie es als einzige wagt, die Wahrheit zu sagen.

H. F., Jüterbog

. . . . Unser Herr B. ist übrigens seit langen Jahren Abonnent der „Motor-Kritik" und dieser hat sie bereits weitestgehend empfohlen. Wir können Ihnen für den stets aufrichtigen Inhalt nur unsere größte Anerkennung aussprechen.

„Traktor und Anhänger"

Ich glaube bestimmt, Ihnen in der nächsten Zeit noch mehrere Interessenten für Ihre Zeitschrift aufgeben zu können, und tue dies mit wirklicher Freude, da ich die sachlichen Ausführungen Ihrer Kritiken in unserer Branche für unbedingt nötig erachte, sowie ich mich immer freue, wenn einer den Mut hat, die in der Autoindustrie bestehenden Schlampereien und Mängel zu kritisieren.

W. v. G., Stuttgart.

DIESE URTEILE SPRECHEN FUR SICH

BESTELLZETTEL

An den Verlag „Motor-Kritik" • Frankfurt am Main • Blücherstraße 20/22

SENDEN SIE NOCH AN FOLGENDE INTERESSENTEN UNTER/ OHNE BERUFUNG AUF MICH EINE PROBENUMMER:

Ich bitte hiermit um regelmäßige Lieferung durch die Post der monatlich 2mal erscheinenden illustrierten technischen Zeitschrift für das gesamte Kraftfahrwesen

MOTOR-KRITIK

zum Preise von RM 3.60 vierteljährlich zuzüglich Zustellgebühren bis auf Widerruf

(Ort und Datum)

(Wohnung, Straße und Nr.)

(Name — eigenhändige Unterschrift. — Bitte genau und deutlich angeben)

Bezugsbedingungen: Die Lieferung erfolgt durch die Post, die auch die vierteljährlichen Abonnementsgelder von RM 3.60 zuzüglich Zustellgebühren vor Beginn eines jeden Quartals durch den Briefträger einzieht. Etwaige Reklamationen betr. unpünktliche oder ausbleibende Zustellung sind deshalb an das zuständige Postamt zu richten

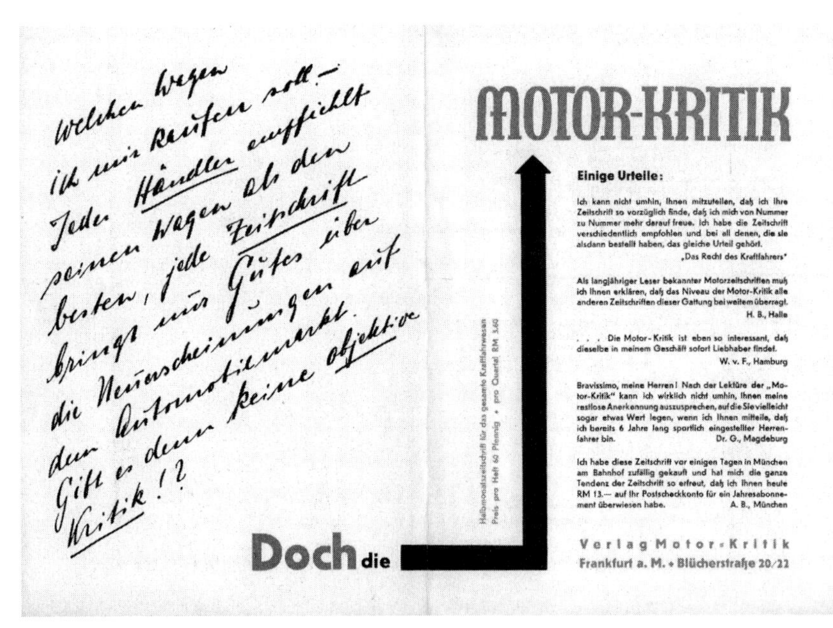

Welchen Wagen ich mir kaufen soll –
Jeder Händler empfiehlt seinen Wagen als den besten, jede Zeitschrift bringt nur Gutes über die Neuerscheinungen auf dem Automobilmarkt.
Gibt es denn keine objektive Kritik!?

Doch die

MOTOR-KRITIK

Halbmonatszeitschrift für das gesamte Kraftfahrwesen
Preis pro Heft 60 Pfennig • pro Quartal RM 3.60

Einige Urteile:

Ich kann nicht umhin, Ihnen mitzuteilen, daß ich Ihre Zeitschrift so vorzüglich finde, daß ich mich von Nummer zu Nummer mehr darauf freue. Ich habe die Zeitschrift verschiedentlich empfohlen und bei all denen, die sie alsdann bestellt haben, das gleiche Urteil gehört.

„Das Recht des Kraftfahrers"

Als langjähriger Leser bekannter Motorzeitschriften muß ich Ihnen erklären, daß das Niveau der Motor-Kritik alle anderen Zeitschriften dieser Gattung bei weitem überragt.

H. B., Halle

. . . Die Motor-Kritik ist eben so interessant, daß dieselbe in meinem Geschäft sofort Liebhaber findet.

W. v. F., Hamburg

Bravissimo, meine Herren! Nach der Lektüre der „Motor-Kritik" kann ich wirklich nicht umhin, Ihnen meine restlose Anerkennung auszusprechen, auf die Sie vielleicht sogar etwas Wert legen, wenn ich Ihnen mitteile, daß ich bereits 6 Jahre lang sportlich eingestellter Herrenfahrer bin.

Dr. G., Magdeburg

Ich habe diese Zeitschrift vor einigen Tagen in München am Bahnhof zufällig gekauft und hat mich die ganze Tendenz der Zeitschrift so erfreut, daß ich Ihnen heute RM 13.— auf Ihr Postscheckkonto für ein Jahresabonnement überwiesen habe.

A. B., München

Verlag Motor-Kritik
Frankfurt a. M. • Blücherstraße 20/22

Top and bottom: Brochure to promote *Motor-Kritik* as the only source of "objective criticism" in the car industry, around 1930.

Opposite page top: *Motor-Kritik* editor Rolf Bielefeld poses to portray the situation of *Motor-Kritik* as a wild dog that the car industry wanted to chain down, 1930.

MOTOR-KRITIK

liest nicht die Motor-Kritik

NUR 1500 MARK

für eine solche
viersitzige Stromlinien-Limousine
mit Schwingachsen

Geschwindigkeit 80 km · Benzinverbrauch 6 L/100 km

werden Sie auf der Automobil-Ausstellung 1933 bezahlen, wenn die Aktionen der Motorkritik von Erfolg gekrönt sind. Lesen Sie das Blatt, das sich zum Ziel gesetzt hat, dem Automobilismus neue, bessere Wege zu weisen, von Mißständen zu befreien und zu höherer Entwicklung zu bringen! · Werden auch Sie Abonnent der „Motor-Kritik“ und stärken Sie damit die Stoßkraft dieses Organes der fortschrittlichen Kraftfahrtechniker. · Gleichzeitig unterrichtet Sie die „Motor-Kritik“ am besten über alle Vorgänge in Auto-Technik-, -Wirtschaft und -Verkehr, denn nirgends sonst wird darüber so offen gesprochen wie hier. Zum Beweis: **Unter RDA-Boykott.**

Above: Leaflet to promote *Motor-Kritik*, stating (below cow) "She doesn't read *Motor-Kritik*" and promising readers that they will soon pay "only 1,500 Mark" for "such a four-seater streamlined limousine with swing axles" if "the campaigns by *Motor-Kritik* are a success." Josef Ganz distributed these leaflets at the 1931 Berlin motor show.

Top left: Josef Ganz (left) and Walter Ostwald, editor of *Motor-Kritik*, enjoying a drink on a terrace in Paris during their visit to the city's motor show, 1933.
Top right: *Motor-Kritik* with a special feature about the Paris motor show, 1933.
Bottom: Rolf Bielefeld, editor of *Motor-Kritik*, demonstrates his design of a system that allowed a sidecar to lean into bends along with the bike, 1928.

Paul Georg Ehrhardt, editor of *Motor-Kritik*, leans against a Röhr, 1930. One year later, Ehrhardt would leave *Motor-Kritik* and become the archenemy of Josef Ganz.

X. JAHRGANG NR. 8 MITTE APRIL 1930

MOTOR-KRITIK

VERLAG FRANKFURT A. M., BLÜCHERSTRASSE 20/22. EINZELHEFT RM 0.60

Moderne Technik

Top left: Josef Ganz (in front) inspects a Röhr Cabrio-Limousine that had just returned from the Monte Carlo rally, before taking it on a test drive, 1930.

Top right: *Motor-Kritik* featuring the test drive of the Röhr Cabrio-Limousine, 1930.

Bottom: The Röhr Cabrio-Limousine after colliding with a tree head on, just as Josef Ganz was bending down under the dashboard to open the valve on the reserve fuel tank, 1930.

Josef Ganz shortly after the accident with the Röhr Cabrio-Limousine that left him with a serious concussion, lacerations to his forehead, and a black eye, 1930.

Top: Ganz in between an old-fashioned Mercedes-Benz (left) and an Adler after having just been hired by both companies as a technical consultant to introduce innovative ideas, 1930. *Bottom left*: The advantages of swinging axles on this Röhr are demonstrated by driving it backwards up some stairs, Darmstadt, 1929. Unfortunately, it caused the radiator to overflow. *Bottom right*: The Stoewer V5 was a small car that, like the Röhr, used front-wheel drive and swinging axles, 1931.

XI. JAHRGANG NR. 11 ANFANG Juni 1931

MOTOR-KRITIK

VERLAG FRANKFURT A. M., BLÜCHERSTRASSE 20/22 EINZELHEFT RM 0.60

Der Fronttrieb in seinem Element

Top: A rear-wheel-driven Adler is compared to a front-wheel-driven Stoewer (not pictured) for *Motor-Kritik* by accelerating both cars up a very muddy country lane, 1931.
Bottom left: Motor-Kritik with a test drive of the front-wheel-driven Stoewer V5, 1931.
Bottom right: Engineer Georg Hoffmann, who would soon join the Stoewer company, after having accompanied Josef Ganz on his test drive with the Stoewer V5, 1931.

Top: One of two prototypes of the Mercedes-Benz 170 during a test drive in the Swiss Alps, 1931. The photo was taken by Josef Ganz on the Furkapas in front of the Hotel Belvédère, next to a giant glacier.
Bottom: Madeleine Paqué, the girlfriend of Josef Ganz, in one of the Mercedes-Benz 170 prototypes after returning from the test drive in Switzerland, 1931.

XI. JAHRGANG NR. 3 ANFANG FEBRUAR 1931

MOTOR-KRITIK

VERLAG FRANKFURT A. M. BLÜCHERSTRASSE 20/22 EINZELHEFT RM 0.60

DIE SCHWINGACHSE
DES
KLEINEN

Top: Josef Ganz (fifth from the left) during the official press introduction of the new Mercedes-Benz 170, 1931.

Bottom left: *Motor-Kritik* emphasising the problems with the swinging axle design of the new BMW, 1931.

Middle right: A BMW prototype after a crash caused, according to Josef Ganz, by its unrealiable swinging axle design, 1931.

Bottom right: The prototype of the new BMW AM1 during a test drive, 1932.

Top: Josef Ganz demonstrates the advantages of his Tatra's swinging rear half-axles by driving it backwards up some stairs, Darmstadt, 1929.
Bottom: Josef Ganz joins Frits Ostwald (behind the wheel), the son of *Motor-Kritik* editor Walter Ostwald, for a test drive in his self-built three-wheeled prototype, 1930. The car had front-wheel drive, steering via the single rear wheel, and a suspension system that consisted of huge "balloon tires."

journal *Stoewer-Magazin* in early 1929, whose aim, Josef Ganz responded in *Motor-Kritik*, was to show "that the swing axle is fashionable nonsense that will disappear with a speed matching the slowness with which it has appeared and that could be advocated only by downright dunces like, for example, us."[11] Stoewer's article achieved nothing. In fact, two years later the company would start marketing swing axles itself.

A second, more direct confrontation followed in September 1929, when the journalist Oskar Weller published an extremely critical article in *Motor-Kritik* following the introduction of a new model from Brennabor, the Juwel (jewel), under the heading "What We Don't Like." "Brennabor has named its new car 'Juwel,'" Weller wrote. "That must mean it's a treasure, a rarity. We hope it remains a rarity. In engineering terms, the car is an antique." Weller explained to readers that, for a start, the center of gravity was far too high and the steering so bad that "one steps into the car with trepidation." He described how it flung itself all over the road and how the back of the vehicle started vibrating and lurching from side to side if the driver braked too hard. He concluded cynically: "A jewel!"[12] The Brennabor directors were furious when they read Weller's comments, and in response they published an article a few weeks later in their company magazine *Brennabor-Roland* entitled "Brennabor Juwel and '*Motor-Kritik*'" that announced in no uncertain terms:

> The publisher is fundamentally an enemy of the German manufacturer and rather than the designers of inexpensive and practical cars finding favor with him, only the tinkerers and innovators do. [...] His advice to the German auto industry consistently serves to make designs more complicated and expensive. [...] The opponents of *Motor-Kritik* should bear this in mind and ask themselves whether they want to become involved with the resulting indirect sabotage of the German economy.[13]

It would not be the last time that Josef Ganz and his magazine were accused by their conservative adversaries of "sabotaging the German economy," despite the fact that his only intention was to use *Motor-Kritik* to furnish the industry with new ideas. Serious personal allegations were made against him and several of his freelance colleagues, and they were maligned in conservative newspapers and periodicals. As well as Oskar Weller and Paul Ehrhardt, Josef Ganz now had several other specialists working for him on a freelance basis, including Rolf Bielefeld, Walter Ostwald, and Frank Arnau. These were the "Angry Young Men" of the car industry, attacking the established order with all the ferocity of anarchist freedom fighters. *Motor-Kritik* was their public platform and Josef Ganz their mentor. Rolf Bielefeld, a redhead with a burly physique, had been taken on to report about motorcycles. Josef had met him only recently, along with motorbike designer Ernst Neumann-Neander, at a demonstration of a system Bielefeld had developed that allowed a sidecar to lean into bends along with the bike. Like Bielefeld, the editor Walter Ostwald, son of the famous chemist and Nobel Prize-winner Wilhelm Ostwald, had practical experience. With his son Frits he had built a prototype, a remarkable three-wheeled car with front-wheel drive, steering via the single rear wheel, and a suspension system that consisted of huge "balloon tires." Ostwald had a quarter of a century of practical experience in the auto and petrochemicals industries behind him, and he had been employed by other specialist magazines, including *Der Motorfahrer* and *Auto-Technik*. As well as writing for *Motor-Kritik*, Frank Arnau, whose real name was Heinrich Schmitt, worked as an adviser to the industry and wrote detective novels. As their final "member of staff," Josef Ganz and his girlfriend Madeleine Paqué took on a large dog, a German shepherd named Dolly that Josef affectionately called "the *Motor-Kritik* scent hound."

No Progress without Criticism

The staff of *Motor-Kritik* all agreed with Ganz that untrammeled, constructive criticism was the only way to achieve their shared objective: the rebirth of the German car industry. Each issue of the magazine featured slogans in heavy type like "Free criticism shortens the path to the goal!" "No progress without criticism!" and "*Motor-Kritik*: contested by stragglers, esteemed by progressives!" Over the course of 1929, subscription applications flooded into the editorial office in quantities matching those of the letters of praise and congratulation. To increase their numbers even further, the publisher set up a competition: the person who managed to enlist the largest number of new subscribers between October 1 and December 10 would receive a prize of a hundred Reichsmark, a considerable sum in those days. There were another four prizes for the runners up, ranging from eighty Reichsmark for second place to twenty for fifth, and twenty consolation prizes of a free six-month subscription.[14] Initiatives like this steadily increased the circulation and influence of the magazine. Ganz used the name *Motor-Kritik* as a quality stamp for the auto industry, turning everything that did not match up to the new progressive standard bright red.

It was a hectic, inspiring, stirring time. Every two weeks, twenty-four times a year, Ganz needed to deliver a complete issue to the printer, with an editorial, articles, photographs, and advertisements. He often wrote in the evenings or at night, in between test drives and visits to car plants. He took many of the photographs himself and sometimes devoted a great deal of time to cropping pictures by hand so that the text would fit pleasingly around them, and he worked to create attractive and powerful cover designs. His front covers featured playful illustrations of the advantages of the innovations he advocated, one example being a picture of a dachshund, with its "independently suspended paws," snuffling in amazement at the solid axles of a capsized toy dog on wheels.[15] Josef Ganz was the driving force behind the magazine's success. He was the face of *Motor-Kritik*, and he was everywhere: attending trade fairs at home and abroad, watching car races, and regularly meeting with auto engineers, manufacturers, and car dealers to test the latest models and prototypes.

In October 1929, Ganz and his colleague Oskar Weller were loaned a marvelous Röhr automobile to drive all the way to the Paris motor show, the Salon de l'Automobile. Covering the well over four hundred miles between Darmstadt and Paris was a real adventure in those days. There were no highways. Only Berlin had a short stretch of freeway, the Automobil-Verkehrs- und Übungs-straße (AVUS), and it was used more for racing than for normal traffic. Along mostly unpaved country roads, Josef, Oskar, and two friends, a pile of suitcases strapped to the back of the car, weaved their way through Germany from village to village, town to town. The Sunday before, Röhr had taken part in a race on the Nürburgring and damaged the mechanical braking system. They were forced to adjust the brakes three times in sixty miles. Late that night, they arrived in Saarbrücken, on the French border, where they had the brakes fixed the next morning. Despite this minor difficulty, Ganz was hugely enthusiastic about the Röhr. The bodywork was conventionally boxy rather than streamlined, but technically the car was the most progressive of its day—a front-wheel-drive vehicle with independent suspension, a strong lightweight structure, and a low center of gravity.

Ganz thoroughly enjoyed his chance to sit at the wheel of this luxurious machine, which unlike practically any other car comfortably absorbed bumps and held firmly to the road. Only a few months earlier, he had published a provocative photo on the front cover of *Motor-Kritik* showing the new Röhr having just climbed backwards up a flight of stone steps. The caption read: "A position that will make the opponents of swing axles, the low-lying chassis, and front-wheel drive think."[16] He had performed the same trick in his own Tatra. That morning, the four men drove from Saarbrücken to the French municipality of Verdun, where one of the bloodiest battles of the First World War had taken place in 1916.

Paris motor show in the Grand Palais

Even thirteen years later, the landscape looked defeated, stripped of its trees and bushes. Many of the fields were still scattered with mines and shells, and they passed huge cemeteries with countless rows of war graves marked by simple white crosses. When they arrived in Paris late that evening, the motor show was closed. They found a hotel and set off the next morning in what was by German standards a busy rush-hour—Paris already had 300,000 cars—for the Salon de l'Automobile in the magnificent Grand Palais on the Champs Elysées.[17]

The annual Paris motor show was a major event to which car manufacturers from all over the world were invited, and the international press came in droves. The Berlin motor show had been cancelled that year because of the poor state of German auto manufacturing, so Paris had taken over as the industry's main European trade fair. All the important car designers and industrialists of the time were there, and Oskar Weller would later write in *Motor-Kritik* about the conversations he and Josef Ganz had with some of them, including Ettore Bugatti, the founder of the flamboyant French company of that name; Ferdinand Porsche, who was then working for the Austrian company Steyr; Hans Ledwinka of Tatra; and Gustav Röhr. Aside from Ledwinka's smaller models, all these designers were concentrating on developing luxury automobiles and racing cars. Weller wrote that Tatra struck him as "progressive without being addicted to innovation, thoroughly sound, including in an economic sense," but Röhr was "actually the most consistent designer of all," applying the most progressive techniques. Weller decided that Tatra and Röhr were the only two carmakers that had managed to combine lower weight with greater solidity. He called Etorre Bugatti "wayward," saying he worked intuitively and built sensational cars. At Steyr, Ferdinand Porsche had developed an extremely large and expensive car called the Austria, using swing axles for the first time in his career. It was based on a design that the Austrian engineer Markus Honsig had developed three years earlier for the Steyr XII. Both Josef Ganz and Oskar Weller were impressed by Porsche's design, but Weller remarked in *Motor-Kritik*: "A shame that these splendid ideas are to be found on such an expensive chassis."[18] Porsche's Steyr Austria was in fact so expensive that shortly afterwards the manufacturer was forced to take the model out of production because of the deteriorating economic climate.

Only a few weeks after the Paris motor show, on October 24, 1929, or "Black Thursday," the global economy suffered a devastating blow. The New York stock exchange fell at such a rate that investors rushed to sell shares, at which point the market collapsed completely. The crash meant that many loans could not be repaid, so banks and businesses folded. Germany was hit particularly hard. Financial aid from the United States ceased, since America now had an immense financial deficit of its own, but Germany still had to make large and extremely long-term war reparations payments to Britain, France, and other countries. After several years of relative stability, Germany quickly sank into a black hole of profound economic depression. Adolf Hitler's Nazi Party seized the opportunity to play upon the Germans' gut feeling that the reparations payments demanded of them were inequitable and the Allies' charge that Germany was responsible for the war unjust.

Crisis

The ailing German car industry was one of many branches of business to be devastated by the financial crisis. Hardly anyone could now afford its expensive and luxurious automobiles. Sales of smaller models declined as well, and the need for better designs and cheaper cars became increasingly acute. For several months, Josef Ganz had been working on new designs. He had applied for a patent on a design for a chassis,[19] and in the weeks before the stock-market crash he sent a development proposal for a Volkswagen to technical director Wittig of Zündapp in Nuremberg, the largest motorcycle manufacturer in the country.[20] For over five years, Zündapp had been looking for a small car to put into production, but its plans had been stymied several times by the poor state of the German economy. Ganz heard nothing from Nuremberg. Because of the crisis, developments and innovations had been put on the back burner right across the industry, and manufacturers were trying to survive by continuing to build the same old standard models. What they needed least of all at this point were revelations in *Motor-Kritik* about how badly designed or dangerous their cars could be.

Partly because of his own motorbike accident in 1923, Josef was greatly interested in road safety, regularly devoting space to it in *Motor-Kritik*. He spent a lot of time on the roads, many of which were of poor quality, and came upon all kinds of accidents, frequently seeing cars that had turned over because their center of gravity was too high, hit trees when the brakes failed, or been abandoned at the roadside with their live axles broken. Safety was a low priority, and accidents could be horrific, with drivers literally impaled on the steering column and passengers thrown out of vehicles, shattering glass as they went. A collision or skid at relatively modest speeds was often fatal. Every time Ganz came upon the scene of a crash, he stopped and took detailed photographs of the wreckage and skid marks. Many of his photos were published in *Motor-Kritik* along with analyses of how the accidents had come about and the technical design faults that were their underlying cause. Even after climbing out injured and bleeding after a crash in early March 1930, he took pictures of the burning wreck of the Röhr Cabrio-Limousine he had been test driving with his colleague Paul Ehrhardt, immediately after its return from the Monte Carlo rally. Ganz had passed the wheel to Ehrhardt, who lost control of the vehicle a few minutes later on a country road near Weimar, driving straight into a tree at the edge of the road just as Ganz was bending down under the dashboard to open the valve on the reserve fuel tank. The car hit the tree head on, at almost thirty miles per hour. All four doors flew open, the engine caught fire, and Ganz hit his head hard against the underside of the dashboard. There was surprisingly little damage to the vehicle, although the tree had moved two inches, roots and all, and Josef was left with serious concussion, lacerations to his forehead, and a black eye.

Ganz later published an article in *Motor-Kritik* about that particular test drive, illustrating it with his photograph of a cloud of black smoke rising from the hood of the Röhr. Despite the seriousness of the accident, Ganz's verdict was decidedly positive. He wrote: "The Röhr as a car? Next to the Tatra the only vehicle with impeccable driving qualities. The minimal unsprung weight and the low center of gravity give it unprecedentedly secure road holding." He did advise Röhr publicly, in *Motor-Kritik*, to use a different steering mechanism and to install more powerful brakes, a reference to the causes of the accident. The contrast between the "phenomenal" Röhr and conventional cars, Ganz said, was immense. "Incomprehensible that so many millions of those unruly devil's carts, in which you always have one foot in the grave, are built and sold every year, now that it has been proven not just theoretically but in practice that vehicles can be made a hundred percent better." All that was necessary, he believed, was to start designing "with the head instead of the ass."[21]

Confrontation with Daimler-Benz

With a severe concussion after his accident in the Röhr, Ganz was prescribed absolute rest at home in Darmstadt. Madeleine Paqué made up a bed in the living room and cared for him lovingly. Friends, acquaintances, and quite a few subscribers sent him get-well messages, but the manufacturer Daimler-Benz took advantage of the accident to open a frontal attack on *Motor-Kritik* with the aim of destroying the magazine. The manufacturers of the Mercedes-Benz had been profoundly irritated for some time by editor-in-chief Josef Ganz's regular criticism of their cars as old-fashioned. With Ganz virtually defenseless, Wilhelm Kissel, director of Daimler-Benz, seized upon a trivial incident as a chance to issue a writ against *Motor-Kritik*, demanding substantial damages. In February, the magazine had described a dramatic reduction in price of the Mercedes-Benz Stuttgart 200 as a "clearance sale." Despite the fact that he had been in touch with Daimler-Benz since then, Josef Ganz heard about the accusation only two weeks or more after the item was published, while he lay helpless in bed. In response, his publisher stapled a sheet of purple paper to the cover of the next issue with the following text: "Dipl.-Ing. Ganz has been injured in an accident as a passenger during a test drive and will not be combat ready for the next fourteen days or so. He requests a brief truce."[22] To no avail. In his sickbed, Ganz received the writ, issued via the district court in Stuttgart, in which the manufacturer demanded a fine of ten thousand Reichsmark for an "unlawful act." The accusation was also published as a full-page advertisement in the magazine *ADAC-Motorwelt*.[23] A bedridden Josef Ganz deployed all his forces to mount a counterattack. With the help of Madeleine Paqué, he publicized the case far and wide in *Motor-Kritik* and together with his colleague Frank Arnau filled no fewer than nine pages on the subject, concluding:

Dipl.-Ing. Ganz
ist als Passagier bei einer Versuchsfahrt verunglückt
und daher etwa 14 Tage kampfunfähig.
Er bittet um einen kurzen Waffenstillstand.

Paper on cover *Motor-Kritik* issue 5, 1930

> It increasingly seems, [...] with a probability verging on certainty, that what Mercedes is after is nothing less than to destroy the independent magazine *Motor-Kritik*, which is gaining in significance by the day. [...] Does Mercedes have an interest in the annihilation of *Motor-Kritik*? Yes! As long as certain company practices continue, it in fact has an extremely urgent interest![24]

Concussed or not, Josef Ganz was not going to let anyone try to destroy his *Motor-Kritik* and get away with it. With the help of Madeleine Paqué, his colleagues, and his publishers, he in turn submitted an official complaint against Daimler-Benz. Tensions rose to such a pitch that Emil Georg von Stauss of Deutsche Bank, a major shareholder in Daimler-Benz, intervened and demanded that Wilhelm Kissel send him the documents about the case. Von Stauss was furious at Kissel's fierce attack on *Motor-Kritik* and ordered him to settle the conflict immediately if he wanted to keep his job. Von Stauss's secretary sided with *Motor-Kritik* and informed Josef Ganz that evening by telephone of what was going on. The next day, Kissel came to Ganz's bedside in Darmstadt—like a "beaten dog," according to Ganz—and said he was willing to do all he could to resolve the misunderstanding.[25] On April 2, 1930, Daimler-Benz and *Motor-Kritik* reached an agreement. Josef Ganz would acknowledge that no "clearance sale" of the Mercedes-Benz Stuttgart 200 was taking place but that serial production was continuing as normal. Daimler-Benz would then take possession of an entire, as yet unpublished print run of the magazine that contained "damaging statements." The company would pay all the costs of a reprint and extra remuneration for staff, as well as the legal fees and telephone charges incurred.[26] Ganz nevertheless had twenty copies of the issue printed and sent them in confidence to his closest colleagues on the staff of the magazine. He regarded this as the end of the matter. He put one of the withheld, unpublished copies in his filing cabinet, where it would remain until it was confiscated three years later in a Gestapo house search as "evidence" in a blackmail case concocted against him.

CHAPTER FOUR

THE BUG IS BORN

By early April 1930, Josef Ganz had more or less recovered from the severe concussion he sustained in the accident with the Röhr Cabrio-Limousine, and he slowly got back to work. The fifteen months since the launch of *Motor-Kritik* had been truly exciting. In its short existence, the controversial magazine had been commended and praised by many and vilified, maligned, and served with writs by others. Its ferocious campaign against conventional auto manufacturers had caused ill feeling while failing to unleash the technical revolution Ganz had so fervently hoped to see. Nor was there any immediate prospect of a Volkswagen going into production. Ganz worked doggedly at his drawing table at home, producing a complete set of new technical drawings for his Volkswagen design.

That spring, having heard nothing more from technical director Wittig, of the motorcycle manufacturer Zündapp in Nuremberg, Ganz sent development proposals for his Volkswagen to Dr. Carl Hahn, the director of DKW, and to Willy Bendit, the director of Ardie. Both motorcycle companies were keen to broaden their selection by offering small, cheap cars. Their sales were declining, and Ganz was convinced that the bike's heyday as a regular mode of transportation had passed. Despite a verbal pledge from Hahn, Ganz took up a concrete offer from Ardie in late June. Director Willy Bendit had offered him all the production facilities at his factory so that he could build a Volkswagen prototype. At that point, Hahn wrote Ganz a letter saying his decision to work with Ardie had come as a "great disappointment" and tried in vain to persuade him to switch to DKW.[1] A technically advanced and successful Volkswagen might achieve a highly competitive position in a market awash with inferior cars. "Unlikely though it may sound, despite there being hundreds of different types to choose from, not a single one is free of extremely serious faults," Ganz wrote that month in *Motor-Kritik*, and: "If one model avoids a particular failing, there are so many other things you simply have to put up with." What Germany badly needed, he said, was:

> A distinctive two- or three-seater. Weight 400-600 kg [roughly 880-1,320 pounds]. 0.75-1.5 liter two- and four-cylinder engines. Low chassis, axle-free front wheel suspension, swing axles at the rear. [...] I believe a particularly large market exists for small vehicles built according to the Tropfen method, in other words with an engine at the back and smooth, relatively streamlined bodywork, where possible produced as a convertible, priced at around 2,000-3,500 Mark.[2]

Test Drive in the Benz Tropfenwagen

Josef Ganz's dream was that a new generation of small, cheap, teardrop-shaped cars would flood the market. He saw the Tropfenwagen as a model for the car of the future—including the Volkswagen—despite the fact that Rumpler's streamlined 1920s drop cars lay rusting, discarded and forgotten, in wrecker's yards. Josef Ganz had even seen one being used as a chicken coop. To his horror, Daimler-Benz, which had ended production of its own version of the drop car in 1926, had sold seven of its eight remaining Benz Tropfenwagens to a wrecker that spring for the paltry sum of 1,050 Reichsmark in total. Ganz had seen the eighth during a visit to the company, abandoned in a corner of the factory in Stuttgart-

Untertürkheim. It was only one of two still in existence; the other had been saved from the consistency by a fellow auto engineer, one Herr Morr from Mannheim, complete with a spare engine, chassis, and other parts, for which he had paid 350 Reichsmark. There was nothing Josef Ganz would have liked better than to test drive this legendary racer, similar to the Rumpler drop car that had played a leading role in Fritz Lang's futuristic 1927 film Metropolis, and he put in a request to technical director Hans Nibel. After their run-in over "undesirable comments" in *Motor-Kritik* earlier that year, Nibel was keen to improve relations with Ganz, and he asked his mechanics to make the Benz Tropfenwagen fit to drive again as soon as possible.

It had been a long time since Josef Ganz had looked forward to a test drive with such enthusiasm. On a beautiful summery June day, the restored Benz Tropfenwagen, an impressive monster that looked rather like a large cigar on wheels, was delivered by truck to Ganz's home address in Darmstadt. After an enthusiastic inspection of the car, Ganz settled into the passenger seat next to the factory's racing driver, Willy Walb, who would show him what the car was capable of. A few minutes later, the streamlined Benz Tropfenwagen was racing along a cobbled road far too bumpy for other cars at up to ninety miles per hour. Walb easily took sharp bends at sixty, and when he tramped the brakes the car stayed perfectly on course. This was unprecedented. The Benz Tropfenwagen drove even better than the Röhr cars Ganz loved so much, with their front-wheel drive and swing axles. As Walb tore on across asphalt, Josef Ganz sat next to him thinking of the Volkswagen and the prototype he was about to build at Ardie. Although far smaller, like the drop car the Volkswagen would have an engine mounted just in front of the rear axle, which in combination with independent suspension would deliver similarly impressive road-holding. Shortly after the test run, the wildly enthusiastic Josef Ganz once again emphasized to Hans Nibel the great importance of the dropcar concept and pleaded with him not to dismiss the idea of using rear-mounted engines and swing axles in future Mercedes-Benz models. Hans Nibel and the engineer Max Wagner had by this point developed a prototype of a new Mercedes-Benz with a solid front axle and swinging rear half-axles that were very similar in design to those used by Tatra,[3] but they were experiencing problems putting the car into production. In the next issue of *Motor-Kritik*, published in early June, Josef Ganz publicly expressed his enthusiasm for the Tropfenwagen:

> The ideal vehicle, a Tropfenwagen like this. [...] In all the world there is surely no other car that has such a grip, such absolutely imperturbable road holding. [...] The test drive in the Benz-Tropfen has once again convinced me that the current way of building cars is a dead end not just economically but above all technically. [...] The Tropfenwagen is destined to become mainly a light two- or three-seater weighing around 400 kilos [880 pounds] empty, which will make it as economical as it is effective and comfortable. And if made even smaller and cheaper it's a Volkswagen.[4]

Ardie Volkswagen

Ganz could still feel the adrenaline coursing through his veins when, several weeks after his astonishing test drive in the Benz Tropfenwagen, he set off with a large folder of technical drawings for the Ardie factory in Nuremberg. He was so enthusiastic about finally being able to build a prototype of his Volkswagen, almost ten years after developing the idea, that despite repeated urging by factory director Willy Bendit, he refused to accept payment for it.[5] After all, he was not doing this for his own benefit but in the interests of the German people. If he succeeded in transforming his theoretical design into a practical reality, it would signify a revolutionary breakthrough both for the auto industry and for the German economy in general. The luxury of car ownership would be brought within the reach of ordinary folk for the first time in history. Willy Bendit had reserved space on the factory floor and a range of machine tools for the building of the prototype. Ganz set to work wearing a long gray dust jacket. Since the motorcycle manufacturer Ardie did not have a fully equipped car factory, the means of production were limited, but Ganz made do with the lathes, milling machines, drills, and hand tools available. He was inventive enough to work with limited facilities, and in any case, the point now was to make an operational prototype, not a production-ready vehicle.

For the main element of the chassis, Ganz used a simple, sturdy, straight steel tube with a circular profile, which formed the car's "backbone." It was a principle inspired by the natural world. He welded the mountings for the suspension at either end, and in the middle a crossbeam for the seats and the floor. These mountings would eventually hold the bodywork in place as well. Ganz placed between the seats and the rear axle the entire propulsion system, including the engine, the gearbox, and the fuel tank. The power was transmitted via the gearbox to the swinging rear half-axles by a chain-drive mechanism.

In early August, Ganz had to leave off developing the Volkswagen at Ardie for a short while to take a brand new six-cylinder Citroën for a test drive in the Swiss Alps. He seized the opportunity to visit his sister Margit, who after divorcing the actor Jakob Feldhammer had married Hugo von Tolnai and given birth to a son, also called Hugo. After picking up the Citroën convertible in Darmstadt, Josef Ganz and Madeleine Paqué drove to Worms, where they put their German shepherd Dolly in kennels before traveling on to St. Niklausen to spend time with Margit Ganz in Uncle Alfred Ganz's Villa Solina. Josef Ganz was very fond of his two-month-old nephew, who could cry so loudly that he suspected "Mussolini would be envious of his voice." After a pleasant few days at the Villa Solina, Josef and Madeleine set off for the Swiss Alps, taking a route on which Ganz was "familiar with every rock" from his earlier test drives in the Hanomag and the Dixi. In his article in *Motor-Kritik* about his experiences with the Citroën, Ganz hinted at the development of the Ardie-Ganz prototype, telling his readers that perhaps he would come back to the Swiss Alps the following year in a small car that would be even more "lightweight, cheap, and economical" than the Hanomag. "Strictly confidential: a design of my own."[6]

The Concept Proves Its Worth

Later that August, after Ganz returned to Germany, the drivable chassis of the Ardie-Ganz prototype was completed. Under the watchful eye of director Willy Bendit and several factory workers, ignoring the large sign saying "Forbidden to start engines in the courtyard," Josef Ganz ignited the little two-stroke motorcycle engine, which came to life with a great deal of noise, belching clouds of blue smoke. He had handed Willy Bendit his camera to take a picture of this historic moment, and Bendit pressed the shutter with perfect timing as Ganz turned out of the factory forecourt onto the street, driving a bare chassis fitted with two seats and four wheels. The resulting picture turned out rather grainy and slightly blurred, but it recorded the birth of the Volkswagen. The wind in his hair, Josef Ganz drove up and down outside the factory gate at speed. The steering and gearshift were fine, and the independent suspension absorbed all the bumps in the cobbled road, just as the Benz Tropfenwagen had done. The only hazard he faced was the possibility of getting caught by the police, since the car did not yet have a license plate. After a short spin, he turned around and drove back across the cobblestones and through the factory gate.

That night in his hotel room in Nuremberg, Ganz could not get to sleep. The brief test drive he had made in the prototype that afternoon had left him deeply impressed. The simple, small, featherweight car drove far better than he had expected, better in fact than practically any other car he had ever tested. He would have liked to take all his adversaries for a ride in it that very night, to convince them he was right: with well-designed independent suspension, a lightweight car could hold solidly to the road. His prototype had opened the way for the arrival of the Volkswagen, a concept that had a great future, especially if the German government decided to support the initiative. Deep in thought, he tossed and turned. Early next morning, he went back to Ardie to continue working on the prototype. During the lunch break, there was a sudden commotion. While Josef Ganz and the factory hands were busy elsewhere, one of the mechanics saw a man hanging around near the

Ardie-Ganz prototype. The mechanic watched from a distance as he scribbled something into a notebook and made quick sketches of the construction drawings they were using. His suspicions aroused, the mechanic walked over to the man, who ran across the courtyard, out onto the street, and was gone. By pure chance, when one of the *Motor-Kritik* editors, Rolf Bielefeld, returned from a visit to the Zündapp motorcycle factory not long afterwards, Ganz found out who had gotten hold of his drawings. Bielefeld told him that after touring the Zündapp factory he had looked at the new models and then sat talking informally for a while in technical director Wittig's office. A man walked in and introduced himself as Herr Krottenthaler. Unaware of the identity of the visitor, Krottenthaler handed Wittig his notes and the sketches of the Ardie-Ganz prototype in Bielefeld's presence.[7] Bielefeld knew the story of the mysterious infiltrator at Ardie, and he was also aware that Zündapp had approached Ganz again recently about his Volkswagen idea. Because of his commitment to Ardie, Ganz had not been able to enter into negotiations with Zündapp. One question remained: What was Zündapp planning to do with the sketches of the Ardie-Ganz prototype?

Test Drives in Nuremberg

In late August, with the help of the engineers at Ardie, Ganz welded together a simple "bug-shaped" open car body in sheet metal. It was a crude version of a design he had created almost eight years before. The car had one central headlamp and two smaller lights on the front fenders. The open bodywork served to keep the car cheap by using a minimum of material; Ganz was convinced that the future of the Volkswagen lay in a closed, lockable version.[8] In the summer of 1930, fitted with manufacturer license plate IIN 0107, Josef Ganz took it on countless test drives, along with others including Willy Bendit and fellow *Motor-Kritik* journalist Paul Ehrhardt. Throughout those final months, Ehrhardt called on Ganz frequently and followed the development of the Volkswagen with great interest. The two men had endless debates about auto engineering, Ehrhardt repeatedly insisting that front-wheel drive was the only satisfying solution.[9] The truth was that Ehrhardt had a significant interest in passing on details of Ganz's design to his employer, Tatra. Like Willy Bendit and Josef Ganz himself, Paul Ehrhardt had been amazed by the properties of the Ardie-Ganz prototype. The suspension comfortably absorbed the vibrations caused by the cobbled streets of the ancient city of Nuremberg, which a year earlier had served as the backdrop to a mass Nazi Party gathering. Adolf Hitler's movement was gaining in influence once again as a result of the continuing economic depression.

In contrast to rival political parties, which were wracked by internal bickering, Hitler's Nazis had a powerful, unambiguous message: the economic crisis was the fault of Jewish financiers and Bolsheviks. In elections to the Reichstag on September 14, 1930, the Nazi Party won 18.3 percent of the vote, making it the second-largest party in Germany. Josef Ganz had little time to absorb the news of this alarming election result. That day, his full attention was taken up by the need to complete the forthcoming issue of *Motor-Kritik,* due to be delivered to the printer the following morning. In it, he presented the Ardie-Ganz prototype to his readers for the first time. The cover featured a picture of Ganz sitting proudly at the wheel of his prototype as he took it for a test drive in the forests around Nuremberg. In the accompanying article, "The Ardie Experimental Minicar 1930," he wrote:

> All the structural details are still missing, since up to now it's been purely a matter of determining whether a vehicle as light as this can be made sufficiently steerable and shockproof. The test drives carried out so far with this totally anomalous construction have not only proven that it's possible to build a lightweight vehicle that's absolutely safe to drive but that amazing results can be achieved, with absolutely solid road holding of a kind quite unknown until now. The high natural oscillation rate of the suspension system means the wheels never bounce free of the road surface, even on corduroy roads. The car remains steerable with one finger, able to describe a zigzag pattern effortlessly.[10]

Worldwide Success

The news published in *Motor-Kritik* about the Volkswagen built by Josef Ganz at Ardie swept through the industry like a tornado, but it was not received with universal enthusiasm. Many manufacturers regarded a cheap Volkswagen with better driving qualities than their own conventional models as a serious threat—and in his influential magazine Josef Ganz was determined to promote only the most technically progressive models. Since his appointment as editor-in-chief of *Klein-Motor-Sport,* Ganz had transformed a magazine with just three hundred subscribers into the renowned but highly controversial specialist journal *Motor-Kritik,* with a print run of thirteen thousand copies and a readership in thirty-five countries.[11] In acknowledgment of his success, the publisher raised his salary considerably in November 1930, from seven hundred to a thousand Reichsmark per month. This further expanded the opportunities open to him.

The successful test drives in the Ardie-Ganz prototype had made Willy Bendit extremely interested in developing the car further to create a production model. If he could then position it as an affordable Volkswagen, he would have seized an extremely competitive position for himself—which was exactly what established car manufacturers feared. In consultation with Josef Ganz, the decision was made to give it more conventional-looking bodywork instead of the open streamlined design, something better suited to the overwhelmingly conventional target market. Although Ganz was

Motor-Kritik's increasing readership, 1931

a passionate advocate of streamlining, he did not feel it was absolutely essential in a small open Volkswagen with a maximum speed not much in excess of thirty-five miles an hour. The new bodywork had a thoroughly conventional appearance, complete with a fake radiator at the front, a vertical front windshield, and a rear end that tapered to a single point, the so-called boat tail that was used for sports cars.

Legal Amendment

For Bendit, there was one essential prerequisite, without which he could not start production. Legislation would have to be amended so that the Volkswagen could be driven with a motorcycle license. Motorcyclists were after all the largest target group. Under the existing traffic laws in Germany, an exception was made only for cars with three wheels, a maximum engine capacity of 0.2 liters, and a maximum weight of 350 kilograms (772 pounds). For years, Josef Ganz had advocated taking only the engine capacity or the weight as criteria, not the number of wheels. Along with Carl Hahn of DKW, who since his abortive negotiations with Ganz had independently developed a small front-wheel-drive car, Bendit filed a proposal with the auto industry's trade association, the RDA, for legislation that redrafted the limit to read simply 350 kilograms. The RDA presented the proposal to the Manufacturers' Commission, a sluggish institution that already had reservations about Josef Ganz and his *Motor-Kritik*. The outcome of the meeting was negative; the manufacturers Opel and Brennabor obstructed the proposed change to the law. Both manufacturers were building small cars according to conventional designs that weighed far more than 350 kilos. Competition from new, innovative, lightweight vehicles might be fatal. Without new legislation, the risk was too great for Ardie to take, and Willy Bendit was forced to abandon plans to put the car into production.[12] This was the second major blow Josef Ganz had suffered at the hands of the RDA in little over a year. "The industry can congratulate itself on protecting its own interests," he wrote resentfully in *Motor-Kritik*.[13]

Nevertheless, in late November 1930, with the best possible references from Ardie, Josef Ganz managed to interest the Frankfurt car manufacturer Adler in his Volkswagen. Although Adler built conventional cars with live axles, the factory was the birthplace of the swing axle that Ganz loved so much, developed and patented by Edmund Rumpler in 1903 when he was working for Adler. In *Motor-Kritik* earlier that fall, Josef Ganz had expressed fierce criticism of the new Adler models 6 and 8, developed by Walter Gropius, the famous German architect and founder of the Bauhaus. In its press release, Adler had praised the shape of the new models as expressing the "harmony of the entire organism," which emerged "logically from its function." Josef Ganz responded in *Motor-Kritik* that although it was a good plan "to mount the bodywork [...] on a conventional chassis," the concept behind the car was fundamentally flawed. "The ways in which we can arrive at types of automobiles that are both aesthetically and structurally satisfying," he wrote, "were indicated by Rumpler and Jaray in 1922." According to Ganz, all the technical and design elements of the "car of tomorrow" were freely available, and he demonstrated his vision with a drawing of a streamlined model featuring a descending roofline and a mid-mounted engine. It was the direct opposite of Adler's latest creations.[14]

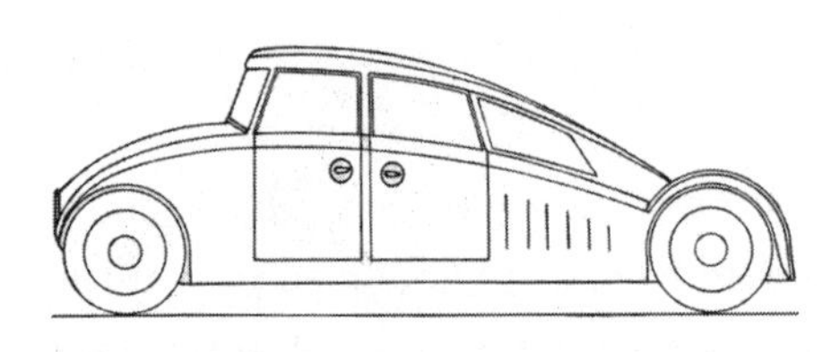

Streamlined car design by Josef Ganz, 1930

Despite this criticism, the directors of Adler, in contrast to most other manufacturers, were open to Ganz's ideas and suggestions. In December 1930, they offered him the position of technical consultant to the factory for a monthly fee of 250 Reichsmark. His main job would be to draw the company's attention to the latest developments in car building and give advice about them—as far as his duty of confidentiality as editor-in-chief of *Motor-Kritik* allowed, of course. He made sure when the contract was drawn up that his work would remain "free and independent" and that "no influence at all" could be exerted upon him, either as an editor and critic or as a confidential adviser to other companies.[15] His first major task for Adler was to build a new, improved Volkswagen prototype. He first developed and patented a better steering mechanism, with two separate steering control rods, one for each front wheel.[16] A second innovation, for which he also applied for a patent, was a setup that allowed the pedals to move horizontally instead of vertically, so that the floor pan, unlike those of other cars of the time, could be sealed against the draft.[17] Based on his design and taking into account his experience with the earlier Ardie-Ganz prototype, Josef Ganz now had a chance to build a second, more presentable experimental vehicle, and in Adler he had the backing of a large manufacturer. At his drawing table back home in Darmstadt, in his white dust jacket, he made a completely new set of drawings.

Hans Nibel of Daimler-Benz was rather dismayed when he heard that Josef Ganz had been taken on as a technical consultant by his rival Adler. He and Wilhelm Kissel had built up a good relationship with Ganz over the past six months and thought he would at least consider taking such a position at Daimler-Benz before signing up with anyone else. Seeking

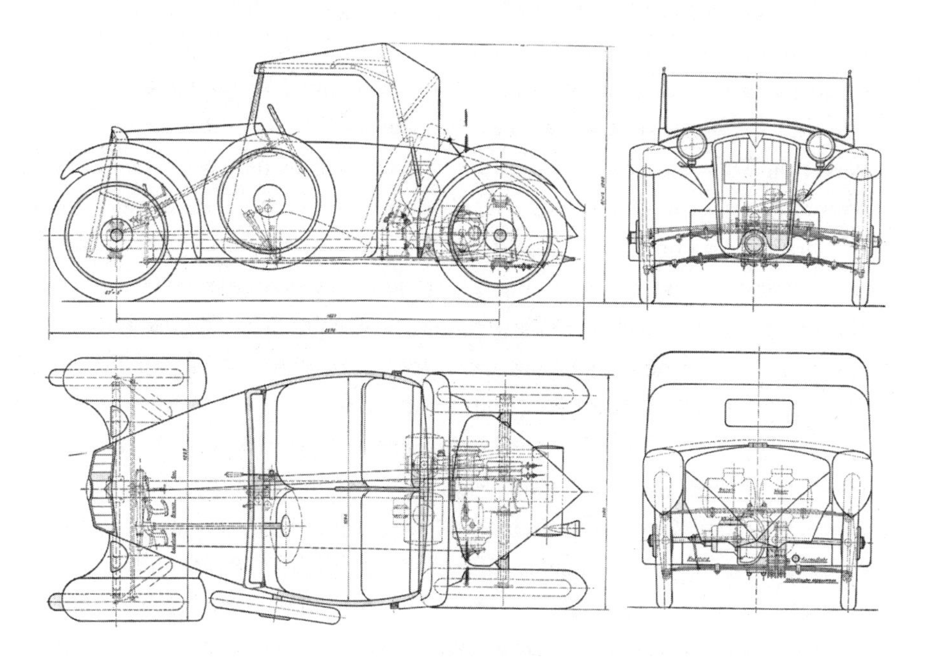

Volkswagen design for Adler by Josef Ganz, 1931

clarification, Hans Nibel rang Ganz that afternoon and invited him to a meeting in Stuttgart-Untertürkheim.[18] During their discussions, which took place a few days later, Nibel told Ganz that he was genuinely interested in his progressive ideas and offered him a position as technical consultant to Daimler-Benz. He could start immediately, in January 1931, for a monthly fee of 500 Reichsmark—double what Adler was paying. In his letter of confirmation, Nibel wrote that Ganz's main task would be to "draw the attention of [Daimler-Benz] to generally interesting issues and phenomena in car building" and, as far as his duty of confidentially allowed, to give "advice on relevant questions as they arise."[19]

Only six months earlier, in the summer of 1930, Daimler-Benz had rejected an offer from Ferdinand Porsche to return to the company[20]—despite messages from Porsche saying that at Steyr, as well as developing a luxury model, he had completely solved "the problem of producing a cheap car for the masses in appropriate numbers."[21] In fact, Daimler-Benz wanted nothing more to do with Porsche, having accused him of funneling company secrets to competitors "in an illegal manner" via his design bureau. As a security precaution, the directors of Daimler-Benz had decided that any employee who was in personal contact with Porsche—even through a spouse—would be sacked on the spot.[22] The first topic of conversation between Hans Nibel and Josef Ganz in his new post was independent suspension. Ganz told Nibel about the exciting test results with the Ardie-Ganz prototype and in mid-December sent him a photograph of its rear axle and an independent article about suspension systems of the same general kind by his journalistic colleague Max Troesch of the specialist magazine *Allgemeine Automobilzeitung*, who had reached an extremely positive verdict on a comparable concept.[23]

Despite the failure to secure commercial production of the Ardie-Ganz prototype, Josef Ganz and Madeleine Paqué toasted the arrival of 1931 in high spirits. *Motor-Kritik* was becoming more successful by the day, Ganz had been given a pay raise by his publishers in November, and now he held two positions as technical consultant. In January, he would start building a whole new prototype of the Volkswagen, this time at Adler. That New Year's Eve, the future looked exceedingly bright.

CHAPTER FIVE

SABOTAGED BUG

In those first cold winter months of 1931, Josef Ganz traveled between his house in Darmstadt and the Adler factory in Frankfurt several times a week, along streets and country roads covered in deep snow. Its swing axles enabled his Tatra to perform well in such conditions, whereas many conventional cars would skid if the surface was even slightly slippery. The number of crashes in those years was limited, but only by the fact that there was so little traffic on the road. At Adler, Ganz could often park his Tatra directly in front of the entrance to the nineteenth-century main building, above which, in a semicircle decorated with two lion's heads carved in stone, stood the name of the factory and its founder: "ADLERWERKE—HEINRICH KLEYER AG" Dr.-Ing. Heinrich Kleyer had started producing bicycles there in 1880 and then progressed to cars, motorcycles, and typewriters. Fifty years later, the factory was still operating under the watchful eye of its founder, now seventy-seven years old.

As far as production technology went, Adler was a fairly modern factory, with a large quantity of machine tools and its own power plant. It had partially switched to production-line methods, but a car's chassis was still pushed by hand from one assembly shop to the next until finally the bodywork, built elsewhere in the factory, was affixed. The most modern car factories in Germany were those of the American Ford company in Cologne and Opel in Rüsselsheim, which came into full American ownership in 1931 when it was bought by General Motors. Josef Ganz visited both factories that year and watched how cars were built using assembly-line manufacture, in accordance with modern production principles imported from America. At Opel, he got to know sales manager Heinrich Nordhoff and built up a good relationship with him. The models produced by Opel and Ford, like those of Adler, were reasonably sound but conventional designs with none of the technical features that Ganz longed to incorporate.

Building the Prototype

Ganz set to work at Adler in an area set aside for him in one of the buildings on the factory site, which covered a hundred thousand square meters in total and was divided into four production complexes. In the middle of the vast, chilly shop floor, he created a setup with two large wooden supports on which he could construct his chassis at a suitably ergonomic working height. As with the earlier Ardie-Ganz prototype, he used a straight steel tube as the basic element of the backbone chassis, to which he welded mountings front and rear for the leaf-spring suspension. At the front was his patented steering system and at the rear the engine and gearbox, with chain transmission to the rear half-axles. The track width was greater at the front than at the back, because there was no differential between the swing axles of the two drive wheels at the rear. The only other attachments to the central tube were the two crossbeams that would hold the leather-clad seats, the corrugated metal floor, and the bodywork. After several weeks of diligent construction work in the cold Adler factory, Ganz put the finishing touches to the chassis of his new prototype in the spring. The design was astoundingly simple and effective.

Adler gave the chassis the manufacturer license plate IT-0444, so that Ganz could take it for early test drives in the streets around the factory site. The car was really quite dangerously small and low for driving between cars that stood so much higher on their wheels. In case of emergency, he fixed a brass hand klaxon with a large rubber squeeze bulb to the steering column. To save weight, the little car had no electric starter motor but instead a mechanical device with a lever. Josef Ganz sat at the wheel of his latest creation and yanked the starting handle. The two-stroke engine sprang into life. He stepped on the clutch and pushed the gear lever, which stuck out between the two seats from behind, pointing it upwards and to the left. Revving the engine, he released the clutch, and the car quickly began to accelerate. He changed into second, then into third, and tore past the Adler factory buildings at a top speed of about thirty-seven miles per hour. Refusing to slow down for bends, trying to hit as many bumps in the road as he could, and tramping the brakes, he tested whether the prototype could be thrown off course. It was a truly magnificent car, he concluded. In fact, it was what he had always intended it to be: a scaled-down and improved mini-Tropfenwagen.

Campaign for the Streamlined Car

The only thing the new prototype lacked was streamlined bodywork. Ever since the development of the Ardie-Ganz prototype, Ganz had been convinced that his Volkswagen would initially have to be put on the market with "open bodywork, little different from the norm and satisfyingly tasteful." It was not just a matter of choosing the cheapest option. He felt that bringing out a car with both a new type of chassis and streamlined bodywork would be to invite "certain difficulties." There was also a practical problem because at that time small engines were still rather noisy, especially in the confined space of a small car. He was hoping to produce a closed version with "ideal streamlining" within one or two years, once the first model was "established and popular."[1] In preparation, he had already handed out pamphlets at the 1931 Berlin motor show, now called the Internationale Automobil-Ausstellung (IAA), promoting both *Motor-Kritik* and the streamlined car of the future. One side featured a provocative photograph of a dumb-looking cow with the caption: "SHE doesn't read *Motor-Kritik*." On the other side was a drawing of a small streamlined car with the text:

> Only 1,500 Mark will get you one such four-seater streamlined limousine with swing axles, speed 80 km [50 mph], gas consumption 6 L./100 km [38.75 mpg], at the 1933 automobile show, if the campaigns by *Motor-Kritik* are a success.[2]

That spring, Josef Ganz came up with an alternative bodywork design that was the direct opposite of the "ideal streamlining" he would have preferred. Like the Ardie-Ganz prototype, this little open-topped two-seater had a fairly conventional appearance, including a fake radiator, but its innovative chassis made it extremely low slung. To study the shape, he carefully built an exact scale model out of wood. Dieter Klüpfel, Madeleine Paqué's one-year-old nephew, who often visited with his mother Ute Klüpfel, loved playing with it. Adler used the model as the basis for the bodywork of the actual prototype, which was produced in sheet metal at a scale of one to one and mounted on the new chassis. Unlike the model, the car had no doors; the bodywork simply dipped right down near the seats. Josef Ganz lovingly called his latest creation the Maikäfer (German for May bug). The car had been completed in May 1931, the month when the May bugs started to fly, and at Whitsun many Germans would send each other cards with playfully drawn scenes featuring the little insects, which heralded warmer weather. The little car with its pointed rear and its brown color really did look rather like one.

As the May bugs were leaving their nests, Josef Ganz took his Maikäfer out onto the streets of Frankfurt. Everywhere he parked or stopped to refuel, people thronged to see the amazing little creation up close. They were full of admiration for its compactness and amazed to discover that the engine was at the back. The Bug was so low that Ganz, bent double, was even able to drive it under the trailer of a truck, and it was so light that one person could lift the front end off the ground. He was more than happy to let admiring passers-by have a go at that. He told them enthusiastically that any working man, indeed they themselves, would soon be able to buy such a car, since he had personally made it his goal to motorize the German masses. To the average observer, that sounded too good to be true. The streets were practically empty, and car ownership was rare. The milkman, the baker, and the coal merchant still went door to door in their traditional horse-drawn carts. He estimated that the two-seater, open-topped version of his Volkswagen, once it went into serial production, would cost somewhere between seven hundred and eight hundred Reichsmark, an astonishingly low retail price. The nearest thing to it was the Goliath Pionier, introduced at the recent Berlin motor show, a small three-wheeler with a 0.2 liter, two-stroke, rear-mounted engine and a maximum speed of twenty-eight miles per hour, which at RM 1,400 cost around twice as much.[3]

Streamlined Volkswagen

The little Maikäfer performed staggeringly well, and Josef Ganz got up to the most astonishing pranks in it. Grinning proudly, he drove at full speed into roadside ditches and out again or ploughed across open fields. His passengers were thrilled by the May Bug's remarkable road holding, especially for such a small, lightweight vehicle. He regularly heard comments like "this is the all-purpose car of the future."[4] He demonstrated it to an enthusiastic Paul Jaray and discussed with him at length what streamlined bodywork of the kind Jaray believed in would look like when they applied it to future models. Ganz believed the combination would make this the ideal Volkswagen. Jaray wholeheartedly supported his extraordinary initiative, but among engineers and in car magazines the benefits of streamlining were still the subject of fierce debate, with both proponents and opponents utterly convinced they were right. The difference of opinion was so stark that while *Motor-Kritik* continued its crusade for the streamlined car, its rival *Motor und Sport* decided to publish nothing at all on the subject in future. After its old editor-in-chief Frank Martin had "defected" to *Motor-Kritik*, it had been run by the far more traditional Wolfgang von Lengerke and his technical correspondent Dipl.-Ing. E Friedländer, a conservative who said he could not comprehend why "ideas about streamlining are still taken so seriously in relatively reputable circles." In *Motor-Kritik*, Paul Ehrhardt quoted what he regarded as incredible comments that Friedländer had made in a letter dated May 1, 1931, responding cynically: "Take a look at the date! That's to say: May 1, 1931, and not, as you surely might think, April 1, 1931."[5] Meanwhile, *Motor-Kritik* continued its propaganda offensive in support of modern principles of construction, begun back in 1928. The name Maikäfer—the movement's figurehead—quickly became synonymous with the new nonstandard building methods that had produced a lightweight car with a backbone chassis, a rear-mounted engine, and independent suspension with swinging rear half-axles. It was a unique, progressive model that set an example to the industry, and a battle broke out between manufacturers to see who could be the first to put such a car on the market.

Zündapp Volkswagen

In May 1931, a remarkable press release from Nuremberg arrived at the editorial office of *Motor-Kritik*: Zündapp announced that its factory was to close temporarily so that the production line could be reorganized. Director Neumeyer had resorted to this emergency measure, in the middle of what was usually the busiest season for sales, because of a steep drop in orders for motorcycles. He wanted to switch as quickly as possible to the production of a small car—a Volkswagen. After missing his chance to employ Josef Ganz, Neumeyer had contacted Fidelis Böhler, the designer of the Hanomag 2/10 PS Kommissbrot. It seems more than likely that at this point he produced the sketches of the Ardie-Ganz prototype he had acquired by subterfuge and asked Böhler whether he could develop a similar Volkswagen with four seats for Zündapp at the price Josef Ganz had advocated as a maximum: one thousand Reichsmark.[6] The result was a cross between the old Hanomag and the innovative Ardie-Ganz, a small rectangular four-seater with a rear-mounted, two-cylinder, one-liter boxer engine, a backbone chassis, independent suspension, and swing axles. The simple seats with their tubular frames were removable for use as camping chairs—an idea adopted years later in the Citroën 2CV—and the flat lid of the trunk could serve as a camping table. "If the information from our confidential source is correct," ran Ganz's response in *Motor-Kritik*, "then this is a model that's urgently needed."[7] Zündapp had been a little rash with its announcement, however. In tests, Böhler's prototype developed problems with its suspension, an unacceptable degree of vibration in the chassis, and difficulties with the gearbox. Plans to go into production were delayed for months.[8]

One by one, manufacturers were switching to new models featuring the technical innovations *Motor-Kritik* had been advocating for years, which they had initially scorned. Among the first was Stoewer, which despite its earlier attacks on the magazine was now putting new models into production with front-wheel drive and swing axles. Josef Ganz milked Stoewer's earlier criticism for all it was worth by reprinting an article that *Stoewer-Magazin* had published two years earlier, in which the company had tried, "in the most erudite tones," to show that "swing axles and suchlike are complete nonsense." He also teased the "charming editor-in-chief" of *Motor und Sport*, Wolfgang von Lengerke—another of his critics—who was now proudly driving around in a brand-new Stoewer sports car with swing axles.[9] Deploying the entertaining cynicism his readers had come to expect from him, Ganz succeeded in creating yet more ill feeling among people like Von Lengerke and factory director Bernhard Stoewer.

At Adler too a new wind had been blowing since May, and it was not favorable to Josef Ganz. That month, Paul Ehrhardt, one of the editors at *Motor-Kritik*, used his good connections among the directors of Adler to persuade them to appoint his wartime comrade Gustav Röhr as its new technical director. Röhr's own company, Röhr Auto AG, had filed for bankruptcy at the beginning of 1931, and although it had started up again in April under the name Neuen Röhr-Werke, Gustav Röhr had been excluded from the venture on grounds of mismanagement and theft of company property, including technical drawings, components—even a complete prototype.[10] Gustav Röhr took the hard core of his technical staff with him to Adler, where they operated under the leadership of Joseph Dauben, an experienced engineer who had been responsible for Röhr's innovative cars. At Adler, he was now developing new models with front-wheel drive and independent suspension. There was no longer any place there for Josef Ganz's Maikäfer prototype. Partly at the urging of Paul Ehrhardt, one of Gustav Röhr's first acts on arriving at Adler as technical director was to halt development of the May Bug. Shortly after the prototype was completed, Josef Ganz, in Paul Ehrhardt's presence, was summoned to Gustav Röhr's office. Ehrhardt and Röhr told him that although it might be true, as Ganz claimed, that the Maikäfer prototype was "the start of a new development in car building," the car was as yet "in no sense a production-ready vehicle." Ehrhardt told him: "You have proven that it's possible to build such a chassis with excellent driving qualities," but "now the thing has to be adapted to make it ready for production; the most important work has yet to begin. The problem doesn't lie in the chassis but purely in the bodywork, and I doubt whether you'll be able to come up with an adequate solution for that here."[11] In other words, at Adler the future of the Maikäfer was extremely uncertain, and Josef Ganz should count himself lucky that he had completed the car just in time, before Röhr pulled the plug on the project.

Swing Axles at BMW

Ganz was having more success at Daimler-Benz. In early 1931, technical director Hans Nibel worked on developing a new Mercedes-Benz model, following Josef Ganz's advice on the basic design. The front wheels had independent suspension with leaf springs, without the conventional solid axle used on the earlier prototype, and at the rear were swing axles with coil springs based on a design that Ganz had recently described in *Motor-Kritik.* For all this innovation beneath the skin, the Daimler-Benz directors had conventional bodywork in mind, as had their competitors.

BMW (Bayerische Motoren Werke), which had taken over Fahrzeugfabrik Eisenach in 1928, wanted to equip both its models, the Dixi 3/15 DA4 and the Wartburg DA3, with swing axles. It had built a number of prototypes in late 1930, but two had crashed as a result of design faults. Josef Ganz reported on those accidents and publicly suggested BMW should "go back to the drawing board" with its axle designs.[12] At first, his article provoked a strong reaction from BMW, which said this was "completely false" and claimed that the two crashed prototypes had not been intended for the 1931 model year. In fact the latest prototypes with swing axles had successfully covered over hundred thousand kilometers (62,137 miles). Not long after this minor clash, however, the company approached Ganz to ask him to join them as technical consultant. BMW director Franz Josef Popp said that the license agreement between BMW and Austin was due to expire in 1932 and that rather than opting to extend it, BMW had decided to develop a new model of its own, using a more modern design with independent suspension. Ganz told Popp that in his opinion the future lay in the approach he had taken with the revolutionary May Bug prototype, especially if it were to be given streamlined bodywork. Although Popp was interested, he eventually opted for a rather more conventional model that would not be radically different from the old BMW Dixi. In July 1931, the company offered to employ Josef Ganz as a technical consultant for a monthly fee of five hundred Reichsmark.[13]

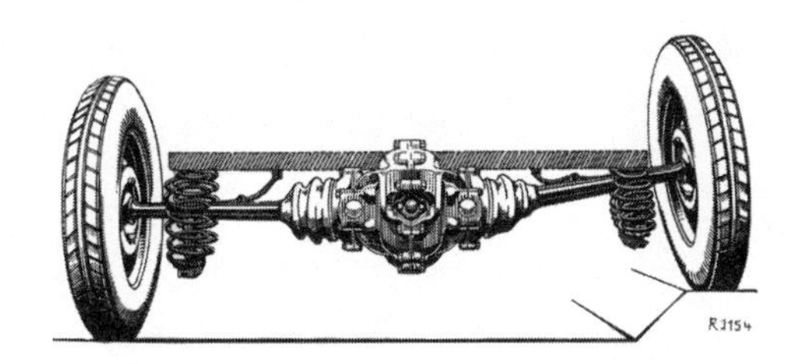

Swing axles of the Mercedes-Benz 170, 1931

CHAPTER SIX

INTRIGUES IN PARIS

Unlike the wrangling in the car industry, the credit crunch did Josef Ganz no harm. He had been taken on as technical consultant by three manufacturers in the space of six months—Adler, Daimler-Benz, and BMW—and his trade magazine *Motor-Kritik* was flourishing. Anyone hoping to survive hard financial times had to innovate, so a subscription to *Motor-Kritik* was of vital importance. In July 1931, nine months after his previous raise of RM 300, the publisher increased his salary again from RM 1,000 to RM 1,200—in the midst of an economic crisis. Ganz's articles for the magazine and his pioneering work as an engineer attracted so much attention at home and abroad that he was appointed editor of the renowned *Automobiltechnisches Handbuch*,[1] and membership secretary Alex Taub of the Society of Automotive Engineers (SAE) in New York personally invited him to join that august organization.[2] Ganz was a spider in the web of innovation, but these were hectic times. In August 1931, to make his work and his private life easier to manage, he moved to Frankfurt, to the large stately villa at Zeppelinallee 85 where Madeleine Paqué lived with her mother Maria, who owned the building. Ganz rented a number of rooms from her for use as living space, an editorial office, and an office for his technical work. He was now living in the city where the publishers of *Motor-Kritik*, as well as the Adler factory, were based, although he still traveled regularly on the magazine's behalf and often visited BMW in Munich and Daimler-Benz in Stuttgart-Untertürkheim. In the absence of a highway network, these were considerable distances, which Ganz liked to cover in his Tatra.

While the Daimler-Benz engineers, with Ganz's assistance, were putting the finishing touches to the new Mercedes-Benz 170, the directors of the company expressed great interest in the little Maikäfer. At their request, the prototype was loaded onto the back of a truck in Frankfurt in August 1931 for the long trip south to Daimler-Benz. The car was so light that on arrival six factory workers were able, under Ganz's supervision, simply to lift it to the ground by hand. He demonstrated the May Bug to the assembled directors of the company and it was then tested for a full day by Hans Nibel, Wilhelm Kissel, Emil Georg von Stauss, the engineer Fritz Nallinger, and a number of other factory engineers.[3] They tore about at top speed over a nearby stretch of wasteland, the wheels almost leaving the ground as they sped across hillocky terrain. The little Bug performed outstandingly during test runs, and the directors were extremely impressed. At that point, however, they were concentrating on getting the new Mercedes-Benz 170 into production.

Test Drive in Switzerland

The Maikäfer had only just arrived back in Frankfurt when Josef Ganz left with an engineer, a mechanic, and a test driver from Daimler-Benz in two prototypes of the Mercedes-Benz 170 for a trip lasting several days. It had been Ganz's idea that the four men should set off from Stuttgart for the unforgiving mountain passes of Switzerland, where over the past few years he had subjected cars including the Hanomag 2/10 PS Kommissbrot, the BMW Dixi, and the Tatra 11 to exhaustive tests. Experience had taught him that this demanding journey through precipitous, snow-covered mountain passes and on cobbled streets inevitably showed up all a car's shortcomings. He greatly enjoyed the scenic drive through Switzerland and took dozens of photographs of the mountain roads, the villages, and the surroundings of the Hotel Belvédère, beautifully situated on the Furka Pass at an altitude not far short of eight thousand feet, with a view of a giant glacier.

One of the greatest challenges was the road over the Stilfersjoch, the highest pass in the Italian and Austrian Alps, famous for its eighty hairpin bends and its changes in altitude, which totaled 11,500 feet over a distance of only thirty-one miles. Josef Ganz was amazed by the ease with which the two Mercedes-Benz prototypes covered the tortuous route. Zigzagging through the steep mountain pass, he caught up with all the other cars attempting the trip, chatting with his co-driver as he did so about photography and even—with one hand on the wheel and one eye on the road—demonstrating how to change the film in his camera. Ganz became almost reckless on the steep, winding roads. These were the kind of antics he usually dared perform only in cars like the Maikäfer, the Tropfenwagen, or one of Röhr's models. In Brunnen in Switzerland, he visited Paul Jaray to show him the prototype. Over the past few years, Ganz and Jaray had become good friends, and they met quite often in Germany or Switzerland. They could discuss auto engineering and streamlined bodywork for hours.

On the way back to Stuttgart, one of the prototypes was subjected to a final, unexpected test. Just past the historic German town of Offenburg, a Brennabor shot out of a side street backwards at full speed and rammed into the left door of the passing Mercedes. It was over before Ganz knew what was happening. His car spun on its axis one and a half times, skidded another twenty yards, and came to rest straddling the road. He stayed in his seat for another a few seconds, let go of the black Bakelite steering wheel, which he had gripped tightly throughout the crash, and stepped out to survey the damage. To the amazement of the other driver, a smile slowly spread across his face. Crash tests had yet to be invented, and Ganz saw every accident as an opportunity to assess how well a car had been built. Once the formalities were completed, Ganz stepped back into the partially crumpled Mercedes-Benz, started the engine, and, to his surprise, was able to drive off immediately.[4] No technical problems emerged, and a few hours later, closely followed by the second prototype, he drove onto the Daimler-Benz factory site. Wilhelm Kissel was extremely grateful for the extensive test drive Josef Ganz had undertaken with the three company employees and was happy to loan him the undamaged prototype. Josef and Madeleine made several trips in it during the last few weeks of the summer.

The prototypes performed perfectly, but during the final development phase an unexpected problem with the gearbox emerged. The drive shaft between the engine and the gears had started to make a howling noise, despite all the effort the engineers had put into ensuring no mechanical stress would be exerted on it. The factory struggled for some time to correct the fault and then presented it to technical consultant Josef Ganz. To the slight irritation of Daimler-Benz's full-time employees, he was able to locate the difficulty. Investigating the gearbox, he concluded that it had in fact been the mechanical stress on the drive shaft in the prototypes that had prevented it from making a noise.[5] With the problem fixed, the new Mercedes-Benz 170 was virtually ready for production by late September—just in time for the annual Salon de l'Automobile in Paris at the end of that month. The continuing economic malaise had led to the cancellation of the Berlin motor show planned for early 1932, so the directors of Daimler-Benz had decided to unveil their new car in France. That fall, the European motor industry gravitated to Paris.

Introduction in Paris

On a mild, summery evening in late September 1931, Josef Ganz stood and stared in awe at the hundred-foot-high letters C-I-T-R-O-Ë-N, each composed of tens of thousands of lightbulbs, with which the French car manufacturer had illuminated the Eiffel Tower. He had arrived in the late evening, and in order to relax after the long train trip he walked through the ever-busy city center, passing the magnificent Grand Palais, which within hours would open its doors for the Salon de l'Automobile. The Champs Elysées was abuzz with activity when he returned the next morning and walked up the stone steps in front of the neoclassical facade to the main entrance. The exhibition complex was packed. Thousands of people swarmed around the hundreds of cars amid a hum of voices, their cigarette and cigar smoke drifting up toward the roof of the immense space. From steel cables slung between the cast-iron beams of the "glass cathedral," beneath a roof of huge glazed domes, hung the names of legendary French and foreign makes such as Bugatti, Hotchkiss, Delage, Minerva, and Adler.

Josef Ganz was in his element walking among the big luxury cars, many of which sported the gleaming, opulent bodywork of master coachbuilders like Saoutchik and Pourtout. Despite the presence of these beautiful creations, the businesslike design of the technically advanced Mercedes-Benz 170 drew the crowds, and Ganz found himself at the center of attention. He, after all, was the spiritual father of the new model, with its independent suspension and swing axles. Fired with enthusiasm, he told visitors to the Mercedes-Benz stand about the car's ingenious design and unprecedented driving comfort—and this was only the beginning of new developments in Germany. He announced in prophetic tones that soon it would be followed by a radical streamlined car featuring a backbone chassis, independent suspension, and a rear-mounted engine. It was a concept nowhere to be found at the show, where only a few of its individual features could be seen on display, such as the backbone chassis and swing axles used by firms including Tatra and Austro-Daimler. A journalist for the Neue Pariser Zeitung who spoke to Ganz called him "a key figure as regards issues of automobile construction."[6]

Like Daimler-Benz, Neuen Röhr-Werke was presenting its latest model at the Paris show. Ganz took a close look at the impressive front-wheel-drive vehicle with independent suspension and discovered to his amazement that the car had rear axles exactly matching a design he had created himself a year earlier on commission from its then director Gustav Röhr, before the old Röhr-Auto AG filed for bankruptcy. Röhr had agreed to pay him half the money the new construction method would save in production costs but later told him it had failed during tests on the prototype and was therefore unusable. At the show, Ganz spoke to Ernst Decker, an ex-colleague of Röhr who was now the director of Neuen

Röhr-Werke, and asked him how long the new design had been in use. Decker answered: "Ever since you created it for Herr Röhr. It amazed me at the time how rudely Röhr lied in your face." Ganz exclaimed: "That's not a lie, that's fraud!"[7] Decker went on to describe all the changes he had been forced to make before it was safe to put the Röhr car on the road. Gustav Röhr had initially refused each and every one of them. They included reinforcing the chassis, adapting the cylinder head on the engine block, and installing more powerful brakes. Decker said that back in the spring of 1929 he had "begged" Röhr to strengthen the brakes, since they were "absolutely inadequate." Röhr simply replied: "There can be no question of increasing the weight."[8] Josef Ganz had experienced the consequences at first hand in March 1930 when, as a passenger in a Röhr Cabrio-Limousine, he hit a tree and suffered severe concussion.

After his frank conversation with Ernst Decker, Ganz continued walking around the show. He ran into a close acquaintance, Hermann Klee, who was on the board of Wanderer. Klee too had disturbing news, this time about Paul Ehrhardt, who earlier that year had been given a job in the advertising department at Wanderer on Ganz's recommendation. Klee said that the company had caught Ehrhardt embezzling a large amount of money from the advertising budget. As a result, he had been sacked on the spot, and the only reason he was still a free man was that Wanderer had wanted to avoid the "undesirable negative publicity" that a trial would bring.[9] Ganz was shocked by both sets of revelations. Gustav Röhr and Paul Ehrhardt were at the show, which made for a tense atmosphere in Paris. Ganz felt increasingly caught up in a web of intrigue, and Gustav Röhr's wife Margaritta was keen to join in, openly taking it upon herself to blacken Ganz's name. *Motor-Kritik*'s, editor Frank Arnau heard her say to members of the public: "That car of Ganz's is worthless; my husband's had to redesign it completely."[10]

Arnau published this quote from Margaritta Röhr in the next issue of *Motor-Kritik*, in an article about the Paris motor show. An engineering friend of Josef Ganz, one Dr. Knappe, who also recalled unpleasant experiences with Margaritta Röhr, responded: "The lines in question typify the extraordinary objectivity of this 'fabulous' woman, who with her intrigues has dug the graves of so many people who once worked honestly and diligently at Röhr Auto AG, wanting only the best for their factory."[11] It was the start of a long personal war between Josef Ganz and both Paul Ehrhardt and Gustav Röhr, as well as the beginning of a technically oriented but vicious battle between advocates of rear-mounted engines with rear-wheel drive and advocates of front-mounted engines with front-wheel drive. Despite their fundamental differences, both sides now had the same point of departure, namely a desire to make the propulsion mechanism more efficient by eliminating the long drive shaft running from front to back. Both camps also swore by independent suspension using the swing axles that Josef Ganz had been promoting for years.

Ode to Mercedes

Several days after facing the many intrigues of the Salon de l'Automobile, Josef Ganz presented the new Mercedes-Benz 170 to readers of *Motor-Kritik*. He began:

> For many years, Mercedes was seen as a stronghold of conservative car building, wedded to standard methods. Naturally, we launched particularly fierce attacks against it. [...] Daimler-Benz has now set off on a new course, and Untertürkheim is about to become the home of modern automobile manufacture.[12]

Wilhelm Kissel, who as director of Daimler-Benz had tried to bring *Motor-Kritik* to its knees only eighteen months before, was hugely enthusiastic about Josef Ganz's work and wrote to him in response to the article:

> In front of me is the latest issue of your *Motor-Kritik*. The whole layout, the thoroughness and professionalism with which you have assessed our new type "170" deserve the highest praise. Now that we have made the new type public and up to now have received only good reports, I feel compelled to express my warmest thanks for your advice during

> the creation of this vehicle and for your support at its introduction. Above all I am very grateful to you for putting yourself forward as godfather to our new child.[13]

In the wake of Kissel's letter, congratulations on the introduction of the Mercedes-Benz 170 poured in. The motorcycle manufacturer Ernst Neumann-Neander, a friend of Ganz, called it a "100% victory." "Now all the others simply have to follow," he wrote. "Mercedes 170 is a European command." He heartily congratulated Ganz on reaching this "pinnacle" but warned that he should take care that "the poverty of old age (what a terrible term), that scourge of artists" did not catch up with him.[14] Willy Bendit of Ardie was another who expressed to Josef Ganz his "sincere admiration and congratulations," writing: "I hope that despite the economic and financial crisis you will now be free of troublesome daily cares once and for all and can pursue the future with greater independence."[15]

In his article on the new Mercedes-Benz 170 in the specialist magazine *Allgemeine Automobilzeitung*, Frank Arnau wrote that it was "impossible to approach the spirit of the new 'Type 170' without remembering the pioneer of the swing axle, the low chassis, and all the modern construction features seen in this new Mercedes-Benz: Dipl.-Ing. Josef Ganz, who has been of valuable service as advisor to the factory in Untertürkheim. Below the article were two photographs of Frank Arnau in the Maikäfer, with the caption: "The new car by Dipl.-Ing. Josef Ganz for around RM 1,000. Weight: 254 kg [560 lb]—Engine: 0.2 and 0.3 liter Rinne engine—speed: 50 to 60 km/hr [31-37 mph]—Production plant not yet entirely certain, perhaps BMW."[16] Along with the many letters of praise and congratulation, there was criticism. It came from just one individual: Paul Ehrhardt.

> I have no doubt that in its driving qualities the new Daimler towers far above what is currently on offer as an "automobile." To me, however, this is merely evidence that even a faulty application of the new construction principles can produce astonishing results, not proof that here the non plus ultra has been created and that it has risen far above

> all criticism. Take it from me that without your advice and cooperation Daimler would still have brought out the car. (If we are to believe what the Daimler engineers say, then everything they used in the building of it was ready and waiting for them.) I'd have liked to see what kind of criticism you'd have come up with then![17]

There was no justification for any such remark. In the past, Josef Ganz had been just as positive if not more so about innovative models from manufacturers such as Tatra, Röhr, and Gatter—each of which had emerged without any help from him. For all his enthusiasm, he certainly did not believe that this car was the non plus ultra, nor that it had "risen far above all criticism." In his article on the Mercedes-Benz 170, he had in fact impressed upon readers that in judging the new model they should bear in mind that it was merely a first step on the road to modern developments that could lead to "the creation of even lighter and then even more streamlined vehicles, with rear-mounted engines."[18] That, in his view, would be the highpoint of car design. But within the trade, the debate over the usefulness of streamlining was far from over.

The streamline pioneer Paul Jaray followed the fierce arguments closely. In October 1931, in a letter to *Motor-Kritik*, he wrote of the controversy: "It also pleases me to see again and again that *Motor-Kritik* functions in some sense as a calming influence amid the clash of opinions and with its clear and systematic appraisals defends a design for which the foundations were laid in my very first patent for a streamlined car, dated September 9, 1921." The debate between Josef Ganz and Paul Jaray focused on the interplay between the shape of the car and the propulsion mechanism. Jaray had originally thought it would make no difference whether the engine was mounted at the front or in the rear, but during one of Jaray's visits to Frankfurt, Ganz had taken him on a test drive in the Maikäfer and succeeded in convincing him that the way the little car was constructed made it ideally suited to optimally streamlined bodywork. Jaray indicated that he now saw "the rear-mounted engine not as a sine qua non but certainly as the most suitable approach to achieving the most accurately streamlined shape possible." In his letter, he warned against the creeping introduction of "pseudo-streamlining" as a fad intended to "satisfy public taste," since "from an aerodynamic point of view, only a virtually impeccable shape will be of any use." As he put it: "The buyer may be fooled; the air won't be."[19]

Even before the official introduction of its Mercedes-Benz 170, Daimler-Benz, with Ganz as consultant, had continued to innovate by developing the streamlined automobile with a rear-mounted engine that Ganz had been promising to create. In the course of 1931, a series of prototypes of a model it called the 120H resulted, a small car with a backbone chassis, a rear-mounted, air-cooled, four-cylinder boxer engine, and independent suspension with swing axles. Daimler-Benz built several variations on the basic model, some of them featuring streamlined bodywork. During its developmental phase, Ganz advised on aspects including the precise structure of the chassis, suspension, and steering.[20] He was also closely involved with a new patent application for a backbone chassis with independent suspension and a boxer engine mounted behind the rear axles.[21] Although it was largely based on his design for the Maikäfer, to his disappointment the directors of Daimler-Benz felt unable to extend to him partial rights to the patent, "on tactical grounds."[22] Hans Nibel wanted to press on and get the first prototype finished, so he could see how the combination of a rear-mounted engine with swinging rear half-axles would perform in practice.[23] Under Nibel, Ganz worked on the 120H with various other people, including Max Wagner, the engineer who had developed the Benz Tropfenwagen in the early 1920s, and the bodywork designer Erwin Kommenda, who left Daimler-Benz in October 1931 to join the design bureau Dr.-Ing. h.c. Ferdinand Porsche GmbH in Stuttgart, set up a little under a year earlier.

Porsche Tests the May Bug

In late September 1931, Ferdinand Porsche had signed a contract with Zündapp to develop a car built along the same lines as the Maikäfer. The aim was to produce a Volkswagen for a retail price of no more than RM

2,000.[24] Because of disappointing results, Zündapp director Fritz Neumeyer decided in the fall to abandon production of the car developed by Fidelis Böhler, but he still wanted to put a Volkswagen on the market. Porsche and Zündapp had been brought together that September by Adolf Rosenberger, a Jewish businessman and racing driver who had supported Porsche financially in setting up his own design bureau and was now its financial director.[25] Rosenberger was responsible for acquiring clients for the new company and in doing so made use of his contacts at Zündapp. The motorcycle manufacturer expressed a willingness to finance not just the development of a Volkswagen but a racing car project that Porsche and Rosenberger had in mind, based on the design of the Benz Tropfenwagen.[26] Rosenberger had raced the Benz Tropfenwagen in the 1920s, and, like Josef Ganz, he was a keen advocate of the drop car.

When Rosenberger introduced Porsche to Zündapp, the details of the new Volkswagen were discussed and laid down in a contract. After his experiences with the Böhler and with the Ardie-Ganz prototype in mind, Neumeyer knew exactly what he wanted. The contract was extremely detailed, with a full two pages of specifications alone. The car would have a water- or air-cooled two-cylinder boxer engine with a maximum cylinder capacity of one liter, capable of producing twenty-three horsepower, rear-mounted, and preferably built as a single unit, with the clutch, the three-speed gearbox, and the rear-axle housing with swing axles. The maximum speed was to be fifty miles an hour, with fuel consumption of at least thirty miles to the gallon, and 240 miles per quart of oil. The two-door bodywork with four seats would have to be "adapted to suit the latest ideas and conducive to sales in relation to other small types of car." The total weight must not exceed six hundred kilograms (1,323 pounds).[27]

Before beginning the new Volkswagen project for Zündapp, Adolf Rosenberger and Ferry Porsche—who worked in his father's design bureau—visited Josef Ganz in Frankfurt in early November to watch a demonstration of the revolutionary Maikäfer prototype.[28-29] They had driven all the way from Stuttgart in the Porsche company car, a Wanderer convertible, for which Porsche's new design firm had developed the conventional chassis in 1930 as one of its first commissions. The unique bodywork—this was the only example ever built—was by the German coachbuilder Reutter. Ferdinand Porsche had decided to keep it for company use. That fall, the Wanderer convertible would appear in Frankfurt several times, as Ferdinand Porsche took the May Bug for test drives over the weeks that followed.[30] Neither fifty-six-year-old Ferdinand Porsche, nor Adolf Rosenberger, nor Ferry Porsche had any experience in designing lightweight models with rear-mounted engines and fully independent suspension, which could produce a very low center of gravity and superb road holding. Following their test drives with the Maikäfer, the Porsche engineers developed their first drawings, which were presented to the directors of Zündapp on November 6, 1931, along with a cost estimate.[31]

CHAPTER SEVEN

NAZI SMEAR CAMPAIGN

By the fall of 1931, the revolutionary Maikäfer prototype had been the subject of a fierce conflict between Josef Ganz and his rivals Paul Ehrhardt and Gustav Röhr for six months. Tensions reached such a pitch at the Paris motor show in October that shortly afterwards Ganz terminated his contract as technical consultant to Adler.[1] He also sacked Ehrhardt as an editor at *Motor-Kritik*[2] and broke off all contact with him and Röhr. Ganz was fortunate that the directors of Adler had allowed him to keep the May Bug. During all those months, he had made no mention in the magazine of his latest creation. In his first, modest report, illustrated with a collage of photographic material, he wrote in early November:

> It's extremely hard for a creative engineer, who is at the same time a critic, to say anything about his own brainchild. [...] After a year and a half of practical work, what is now before us is above all an early prototype, development of which has ceased. The car was built in a few weeks, and it proves that under certain conditions it's possible to make an extremely light vehicle with good road holding and suspension, something that until now has been regarded by practically all the experts as utopian.

He pointed out that despite approval from colleagues, future production of this type of car now depended "on matters that have very little to do with technical issues."[3] In the first week of November 1931, when Josef Ganz published his article about the May Bug, radical Nazi sympathizers were receiving the latest issue of *Die Nationale Front—Kampfblatt für Deutsche Politik und Deutsche Wirtschaft* (*The National Front—Activist Paper for German Politics and German Commerce*). Ganz had never heard of this obscure publication, but a concerned reader drew his attention to a defamatory article that had appeared in it headed "*Motor-Kritik*." He promised to send Ganz a copy of the issue, in which he was maligned as a "Jewish engineer" and his magazine as a "poisonous excrescence." It arrived at the editorial office the next morning with the rest of the mail.

Poisonous Excrescence

The front page of *Die Nationale Front* carried a prominent swastika, and on the next page was a large picture of Adolf Hitler and Hermann Göring, whose radical right-wing Nazi Party was gaining influence as a result of the financial crisis. Josef Ganz quickly leafed through to the article entitled "*Motor-Kritik*" and read:

> Frankfurt am Main has of old been a city of journalistic excrescences. [...] No surprise then that it's here a new poisonous excrescence has emerged over the past few years, "*Motor-Kritik*," led by Herr Josef Ganz, a Jewish engineer who as such has never proven his constructive or productive capacities but has thrown himself into writing for the trade press and now produces a quantity of corrosive criticism that knows no parallel.

The article was indeed highly defamatory, portraying Josef Ganz as an enemy of German industry. The "Ganz method," it said, was to criticize models from one German manufacturer after another, systematically, until the director of each targeted company in turn was forced to pay "hush money" in exchange for advertising space in *Motor-Kritik*. "Another method," the anonymous author went on, "is to appoint the man as the factory's freelance 'technical consultant.' After that they are left in peace, as long as they pay a lot of money, that is." Fellow journalist Frank Arnau—"another man with a destructive tendency"—was dragged through the mud as well. This "Jewish writer" was said to "imagine himself able to save the German auto industry by his writing—although he is easily replaceable."[4]

Having read the article with mounting astonishment, Ganz had a strong suspicion that Paul Ehrhardt and Gustav Röhr were behind it. This particular Nazi magazine was edited by one Hans Köth, a shady character who, like Ehrhardt, had once been accused of fraud. Ganz responded with a brief report in his own magazine: "The enemies of *Motor-Kritik* have now started mixing with company that is absolutely worthy of them."[5] The article was a grave attack on Ganz's professional competence, and he feared it might taint the image of his most important employer after *Motor-Kritik*, Daimler-Benz, which could certainly do without any negative publicity for its recently launched innovation drive. With a heavy heart, Ganz sat at his typewriter and wrote to the directors of Daimler-Benz:

Ad for the Mercedes-Benz 170, 1932

> In the year that I have worked with you I have experienced such an endless number of good things and as a result feel such a bond with you [...] and your factory that it is very painful for me to think that any disadvantage could accrue to you from my collaboration. I therefore request you to regard the contract between us having been returned entirely into your hands. Please regard this as reiterated testimony to how well disposed toward you I have become over the past year.[6]

The Daimler-Benz directors had no desire to end their collaboration with Josef Ganz. The enormous sales success of the Mercedes-Benz 170 was looking like the one thing that might save the company in harsh economic times, and under Ganz's consultancy a further step had been taken with the development of the new prototype 120H, built according to Maikäfer principles. In their answering letter, the directors expressed great appreciation for Josef Ganz's honorable attitude but they categorically rejected his proposal to cancel their contract with him. "Rather we respectfully ask you to continue to stand beside us with your valuable advice."[7] For Ganz, this was extremely good news.

The article in *Die Nationale Front* was only the start of a broader campaign of defamation, as became clear five days later when the new issue of *Motor und Sport* landed on the desks of the *Motor-Kritik* editors. On opening it, Ganz's eye fell on an article headlined, in a large bold typeface: "The Car of Tomorrow?" He read the strange epistle from his rival Wolfgang von Lengerke with utter amazement. It was a venomous response to a number of quotations picked out of an article Ganz had written and published in *Motor-Kritik* six months earlier. Here they were being used as part of an attack on the well-known critic and his propaganda campaign in favor of the streamlined car with a rear-mounted engine. "You are the coming man," Von Lengerke wrote cynically. "Without you the entire industry is a mess." He accused Ganz of "delusions of grandeur." In the article from which the quoted fragments were taken, Ganz had expressed the opinion that although streamlining for cars was starting to take off as a fad, there was as yet little sign of its being applied in ways that were scientifically sound.[8] "What a puritan!" was Von Lengerke's reaction. "How much, dear master, would it have cost us to begin the revolution from scratch every six months? You know how much, right? [...] We're all consistently mistaken except for—why should I be embarrassed to say it—Ganz, Josef, who isn't." According to Von Lengerke, streamlining was relevant only for racing cars, not for ordinary passenger vehicles. "In a hundred years from now maybe... Maybe... As far as that goes I don't want to be a prophet! But maybe not, too." In an attempt to show that streamlined cars with rear-mounted engines were "impractical and ugly," he illustrated his article with two photographs of a Burney Streamliner, a large and relatively cumbrous streamlined British model from the late 1920s with a rear-mounted engine. "Or do you, dear reader, perhaps think the cars pictured here are attractive?" Von Lengerke asked his readers with a sneer. "They're whales on wheels... 'Yes, but functional,' I hear someone say. [...] Aside from that, it's quite a different thing to recommend that the German car industry should build exclusively cars like this—exclusively!"[9] Von Lengerke conveniently neglected to mention that Ganz had in fact been critical of the Burney Streamliner in the past, saying in *Motor-Kritik* that Burney had "missed his target and gone to the opposite extreme." He had offered suggestions for an improved version.[10]

After reading the article, Ganz immediately wrote a response and mailed it that afternoon. Three weeks later, his letter was printed in *Motor und Sport* with Wolfgang von Lengerke's reaction underneath. Like the anonymous author of the defamatory article in *Die Nationale Front*, Von Lengerke took a firm stance against Ganz's dual role as both journalist and technical consultant to the industry. Setting aside the debate about the content of the article, he wrote, "what's much more interesting is that the publisher of *Motor-Kritik* is an engineer who gives advice to the industry [...]. Do you believe that a person can write theater reviews if he is himself a director, or even simply a playwright?"[11] Josef Ganz did not return fire a second time in this magazine war, but nevertheless he received an unpleasant surprise in December 1931 when he saw the Christmas issue of *Motor und Sport*. It contained a transcript of a board meeting at "a large car manufacturer"—recorded by hidden microphone, the magazine claimed, by one of its editors. The article began by saying: "This has given us the opportunity to offer our readers a precise account of the golden words spoken during this meeting and the colossal rescue plans once again drawn up for the German car industry." Although there was no mention of the specific company involved, and all the names had been changed, the article contained barely disguised references to Josef Ganz (fictional name "Dipl.-Ing. Pfau") and his colleague Frank Arnau (fictional name "Krakauer"). The aim was clearly to ridicule their

Motor-Kritik colleague Georg Ising joins Josef Ganz (behind the wheel) in his second Volkswagen prototype, the May Bug, Frankfurt, 1933.

Above: Wooden scale model of the second Volkswagen prototype that Josef Ganz built for the Adler company, Darmstadt, 1931.

Opposite page top: Ganz's German shepherd Dolly next to Dieter Klüpfel, nephew of Ganz's girlfriend Madeleine, who plays with the scale model of the second Volkswagen prototype, Darmstadt, 1931.

Above: Backbone chassis with front suspension of Ganz's second Volkswagen prototype under construction in the Adler factory, Frankfurt, 1931.

Top and opposite page top: The backbone chassis of Ganz's second Volkswagen prototype for Adler with swinging axles and signature mid-mounted engine, Frankfurt, 1931. *Bottom*: Ganz in his second Volkswagen prototype for Adler, nicknamed the May Bug.

Bottom: The "boat tail" of the May Bug, with its mid-mounted engine and swinging rear half-axles, Frankfurt, 1931.

The May Bug was so lightweight that one person could easily lift the front end off the ground, Frankfurt, 1931.

Top: Josef Ganz behind the wheel of the May Bug, showing how incredibly low the car was compared to the cheap small 1920s car behind him, Frankfurt, 1931.
Bottom: The May Bug was considerably lower, smaller, and more lightweight than other contemporary cars, Frankfurt 1931.

Above: Josef Ganz demonstrates the excellent driving qualities of the May Bug before the cameras of German newsreel *Deulig-Tonwoche* in Frankfurt, 1932.
Top right: Josef Ganz demonstrates to the *Deulig-Tonwoche* crew how low the May Bug is by driving underneath the trailer of a truck, Frankfurt, 1932.
Right: Josef Ganz sits inside the May Bug with his colleagues underneath a truck trailer to hide from the rain during filming, Frankfurt, 1932.

Top: The May Bug arrives at the Mercedes-Benz factory on the back of a truck, and factory workers unload the vehicle, Stuttgart, 1931.
Middle: A Mercedes-Benz engineer test drives the May Bug, Stuttgart, 1931.
Bottom: Hans Nibel, technical director of Daimler-Benz, behind the wheel of the May Bug, Stuttgart, 1931.

Middle: Mercedes-Benz 120H prototype.
Bottom: A second streamlined body design for the Mercedes-Benz 120H prototype, 1931.

Top: The Mercedes-Benz 120H prototype with a rear-mounted air-cooled engine, created with the assistance of Ganz and based on the design principles of his May Bug, 1931.
Bottom: Mercedes-Benz 120H prototype with Beetle-like bodywork, Stuttgart, 1931.

Top: The "Volkswagen" prototype designed by Fideles Böhler for Zündapp, 1931. The simple seats with their tubular frames were removable for use as camping chairs. Zündapp, however, decided not to put the car into production.
Bottom left: Josef Ganz continued to test drive the May Bug prototype under all conditions to further improve the design and find a manufacturer, 1932.

JAHRGANG NR. 21 ANFANG NOVEMBER 1931

MOTOR-KRITIK

ERLAG FRANKFURT A. M., BLÜCHERSTRASSE 20/22. EINZELHEFT RM 0.60

Um den Kleinstwagen

Motor-Kritik in which Josef Ganz introduced the May Bug prototype to his readers, 1931.

A blow-up from the *Motor-Kritik* cover reveals Ferry Porsche behind the wheel next to Adolf Rosenberg, also from the Porsche company, test driving the May Bug, Frankfurt, 1931. The Porsche company had just signed a contract to develop a new "Volkswagen" prototype for Zündapp.

After test driving the May Bug, the Porsche company developed a similar backbone chassis with swinging axles and a rear-mounted engine for Zündapp, 1932.

Above: Ferry Porsche next to a prototype body for the Zündapp 12, 1932.
Below: Zündapp12 prototype, 1932.

Top: An engineer of the Standard Fahrzeugfabrik test drives the new chassis of the Standard Superior, a small Volkswagen designed according to the patents of Josef Ganz, 1932.
Middle and left: The prototype of the Standard Superior during a test drive. In the production model the cabrio concept was dropped, 1932.

Top: Brochure for the new Standard Superior, the "safest and most high-end small car," 1933.
Bottom: *Motor-Kritik* showing the chassis of the new Standard Superior, 1933.

Ad by the Standard company in *Motor-Kritik*, claiming "all desires of MK readers fulfilled," 1933. It shows the preproduction model of the Standard Superior as it was introduced at that year's international motor show in Berlin.

98 STANDARD 98
UZ 74
74 DRAUZ 74
DRAUZ

Above: Gustav Rau, the financial director of Standard, at the international motor show in Berlin, 1933.
Left: The chassis of the Standard Superior was on prominent display at the international motor show in Berlin, 1933.

Top and bottom: Josef Ganz has parked the exhibition model of the Standard Superior next to the Zündapp12 prototype that Ferdinand Porsche himself had driven to the international motor show in Berlin, 1933. Zündapp by then had already decided not to produce Porsche's design, so it stayed parked outside. This car was considerably bigger than the Standard and therefore too expensive to become a "Volkswagen."

Various small cars at the 1933 Berlin motor show: the Hercules (top left), and the Theis, with its streamlined body that served as a "sleeping cabin" (top right). The stand of the Neuen Röhr-Werke showed a backbone chassis with swinging axles (bottom left), and the "sensationally cheap" four-seater car by Opel (bottom right).

Brief aus dem Leserkreise.

JAHRGANG NR. 16

KLEIN-MOTORSPORT

MOTOR-K

hinter dem offiziellen rucken der ausstellung!

Reichskanzler Adolf Hitler dankte den Herren der Industrie, den genialen Konstrukteuren und Technikern sowie der großen Armee der deutschen Arbeiter für ihren beharrlichen Arbeitswillen.

lieber herr ganz,da muss ich etwas klarstellen in der rede? - wir wollen doch gerecht sein: ursinus,ihnen,dem maikäfer,den mitarbeitern, kurz der m.-k.mit dem beharrlichen arbeitswillen verdanken viele die gedanken,die 1933 in berlin veranlassung gaben zum danken.

gruss ihr

A controversial and daring collage, published in *Motor-Kritik*, to illustrate a reader's letter stating that we have to thank Josef Ganz (with his May Bug design) and the *Motor-Kritik* team for fostering the innovative designs that were an eminent presence at the 1933 Berlin motor show.

Bird's eye view of the Mercedes-Benz stand at the Paris motor show, 1930.

Josef Ganz was, in the late 1920s and early 1930s, a regular visitor to the Paris motor show.

Above: Joseph Dauben, engineer at Adler, next to a license-built Rosengart Supertraction, Paris, 1933.
Top: André Citroën at the show, 1933.
Top center: The many brands exhibited at the motor show, Paris, 1931.
Middle right: A front-wheel-driven car by Emile Claveau, Paris, 1931.

Ad for Valentine paint, Paris, 1933.

Above: Horse-drawn carts deliver the exhibition models to the Grand Palais, Paris, 1933.

Top right: Scale model of the Peugeot 301, Paris, 1933.

Middle right: Chenard & Walcker, with eccentric streamlined bodywork by Mistral, 1933.

30. Jahrg. — 21. 7. 1933 **ADAC-MOTORWELT** Nummer 29 — Seite 7

Standard - Superior

Die diesjährige Automobil-Ausstellung hatte eine Reihe neuerstandener Dreirad-Kleinstwagen zum Vorschein gebracht, dagegen nur einen Vierrad-Kleinstwagen für Personenbeförderung, nämlich den Standard-Superior. Dieser konstruktiv recht interessante Typ war aus einer von Dipl.-Ing. Jos. Ganz, Frankfurt a. M., stammenden Versuchskonstruktion hervorgegangen, der sich als Vorkämpfer des Kleinstwagens einen Namen gemacht hatte.

Der Standard-Superior ist, als Ganzes betrachtet, ungewöhnlich gut gelungen — was auf die jahrelange Erprobung der erwähnten Versuchskonstruktion zurück-

Abb. 2. Der Standard-Superior-Vierrad-Kleinstwagen (Frontansicht)

zuführen ist — und wird höchst wahrscheinlich die Einbürgerung auch größerer Gebrauchswagentypen mit Hecktriebsatz einleiten. Bekanntlich findet sich ja schon ein zweiter Heckmotor-Gebrauchswagen, konstruiert von Dr.-Ing. Porsche, im Stadium der Fabrikationsvorbereitung.

Wenn der Standard-Superior, als Ganzes betrachtet, als recht gut gelungen erachtet werden muß, so seien nachstehend doch einige Punkte als nicht restlos befriedigend und verbesserungsfähig gekennzeichnet unter gleichzeitiger Betonung, daß sein einziger wirklicher

Abb. 4. Fahrer, Fahrgast und Gepäck sind im Standard-Superior gut untergebracht

Mangel sein zu hoher Lieferpreis ist. Damit sei nicht etwa gesagt, daß der Standard-Superior als zu teuer im absoluten Sinne anzusprechen ist. Sein Lieferpreis von 1620 RM. liegt aber schon so nahe an demjenigen der zwei billigsten Kleinwagen, nämlich des 1,0-L-Opel und der DKW-Reichsklasse, daß die Preisspanne von zirka 300—350 RM. in vielen Fällen nicht ausreichen wird, den Kaufinteressenten für den Standard-Superior zu gewinnen, trotzdem dieser den Vorteil billigerer Führerscheinbeschaffung und geringerer Haltungskosten, insbesondere geringerer stehender Unkosten, für sich in Anspruch nehmen kann. Fraglos würde eine etwas ver-

Abb. 1. Der Standard-Superior-Vierrad-Kleinstwagen (Seitenansicht)

größerte Preisspanne die Einführung dieses einzigen deutschen Vierrad-Kleinstwagens wirklich fortschrittlicher Gestaltung begünstigt haben. Ob sich indessen im Kleinserien-Bau — Kleinserien-Bau, gemessen an den Herstellungsziffern des 1,0-Liter-Opel und der DKW-Reichsklasse — wesentliche Preisabschläge erzielen lassen, sei dahingestellt. Gerade dieser Umstand wird allgemein die an der oberen Grenze ihrer Preisklasse liegenden Kleinstwagentypen stark im Vordringen hindern,

Abb. 3. Die Einstiegtüren des Standard-Superior sind groß und bequem

wenigstens so lange, als sich nicht eine Großfirma des Kleinstwagens erbarmt.

In Fällen, wo die Vorteile geringer Haltungskosten und billigerer Führerscheinbeschaffung entscheiden und ein ausgesprochener Zweisitzer erwünscht ist, wird der Standard-Superior allerdings trotz seines Preises von der Käuferschaft bevorzugt werden.

Die technische Gestaltung des Standard-Superiors ist modern im höchsten Grade, verfügt das Fahrzeug doch

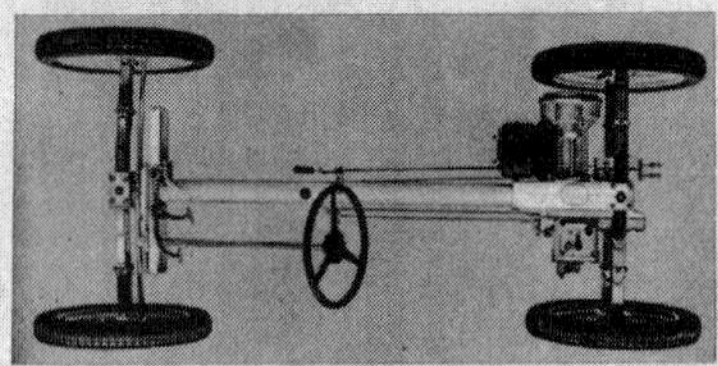

Abb. 5. Das Fahrgestell des Standard-Superior mit Mittelrohr-Rahmen, Front- und Heck-Schwingachse, Hecktriebsatz

The Standard Superior was well received by the press, such as in this first page of an article in *ADAC-Motorwelt*, 1933. According to this publication, the Standard Superior, as devised by Josef Ganz, had "proven extraordinarily successful" and would "lead to the introduction of larger models with rear-mounted engines," such as the one under development by Porsche.

position as consultants to Daimler-Benz. The report began with an opening address by the factory manager, presumably an allusion to Daimler-Benz's director Wilhelm Kissel (fictional name "Steinle"), about a new model for 1932:

> It must be cheap to run and economical; in short, it must as far as possible meet the demands that people have of—and it's true that I'm not keen to utter this overused word—a so-called Volkswagen. Solving this task is made difficult for us by the way certain tendencies have arisen in the automobile market in recent years that have culminated in an aversion to the long tried and tested standard building method.

The tone of the rest of the article implied that "Pfau" and "Krakauer" had no respect at all for the established auto industry and were setting themselves up as its saviors. "Pfau" was quoted as saying: "We expert critics hammer our typewriters to bits, and you so-called car designers build cars that from the outset are ripe for the German Museum!" One "Neumann," an "engineer with twenty years' experience," explains during the meeting that rear-mounted engines are problematic when used in passenger cars. "Pfau" replies: "A competent structural engineer must surely be able to solve that problem easily."[12] Josef Ganz published a response in *Motor-Kritik* in which he explained the point he believed the article was intending to make:

> The publisher of *Motor-Kritik*, a fantastically arrogant, corruptible bully, and a member of the *Motor-Kritik* editorial staff, a garrulous, conceited ass, will be told the truth well and good by the directors of the auto manufacturers (and above all by Herr Friedländer). The point: that everything *Motor-Kritik* stands for—swing axles, the low-lying chassis, lightweight cars, streamlining—is rubbish. [...] To the libelous accusations in this pamphlet we will give what in this case is the only proper response as soon as the anonymous writer can summon the courage to take responsibility for them.[13]

Preis 20 Pfg.

Nummer 12
1.—15. November 1931

Die Nationale Front

Kampfblatt für deutsche Politik und deutsche Wirtschaft

Aus dem Inhalt:

Die nationale Front erscheint 14-tägig und dient:

1. der Stärkung und Förderung des nationalen deutschen Gedankens im In- und Auslande,
2. dem nationalen Ausbau der deutschen Wehrmacht und Landesverteidigung sowie der Heranziehung einer pflichtgetreuen, überparteilichen Polizei und Justiz,
3. der Erziehung einer nationalen, sittlichen und religiösen Jugend, sowie eines vaterlandsbejahenden und pflichtbewußten Arbeiter-, Angestellten-, Beamten- und Kaufmannstandes,
4. der Förderung einer vernünftigen Zoll-, Steuer-, Wirtschafts- und Agrarpolitik,
5. dem rücksichtslosen Kampfe gegen den zersetzenden Marxismus und jede Gottlosenbewegung,
6. dem rücksichtslosen Kampfe gegen Kriegsschuldlüge und Youngplan, Versailler Vertrag, kurz gegen jede Unterwerfungspolitik.

Cover of the issue of *Die Nationale Front* with the article on Josef Ganz and *Motor-Kritik*, 1931

Scapegoat

Just before Christmas, at the same time as the *Motor und Sport* article, a follow-up piece was published in the Nazi magazine *Die Nationale Front.* It said there was "an unexpected degree of interest" in the "fraudulent activities" of Josef Ganz and Frank Arnau at *Motor-Kritik* that had been revealed in the previous issue:

> Countless letters and messages, which flood in to us daily, prove again and again the immense harmfulness of both Galicians Josef Ganz and Frank Arnau. [...] It is in the interests of the German people and the German economy that a clean sweep be made quickly here. We shall therefore pay close attention to this scandalous affair from now on and, in association with the national press, hope soon to be able to put an end to the filthy activities of this Jewish vermin.[14]

The smear campaign was starting to assume an increasingly serious and threatening form, and it threw a dark shadow over the approaching Christmas holiday. Moreover, from his professional colleagues Josef Ganz had heard that the nationwide trade association, the RDA, was once again actively trying to destroy *Motor-Kritik.* It was attempting to establish among its members a list of officially approved test drivers, excluding Josef Ganz and Rolf Bielefeld of *Motor-Kritik.* Ganz made room in the proofs of the next issue for a reaction in which he called on readers who did not believe he and Bielefeld should be excluded to demand to see a test report by *Motor-Kritik* before buying a new car.[15] He was afraid the RDA would use all the power at its disposal to maintain the status quo, thereby seriously jeopardizing the future of the German auto industry. Madeleine Paqué and her mother did their best to decorate the living room cheerfully for Christmas, but Josef spent a great deal of time in his office, either on the telephone or at his typewriter. Between Christmas and New Year's, he contemplated his next move while working hard to get the first 1932 issue of *Motor-Kritik* ready for publication. He took a fresh piece of paper, rolled it into his typewriter, and wrote:

> We look back four years, to when *Motor-Kritik* in its current form was born. The low chassis, swing axles, driving machine, minicars, anti-standard, these are all terms that didn't exist then. And now? [...] Now we've arrived at a point that's been in sight for years. That isn't the fault of those who pushed the cart through the mud but of those who cried "Stop! Stop!" And if in another two years from now foreign countries flood the world with the lightest and cheapest rear-engined, streamlined cars, while we're still sitting in our outdated old bangers, then it'll be just as difficult as it is today to find a scapegoat.[16]

Evidence

Josef Ganz was convinced that Paul Ehrhardt and Gustav Röhr were the driving force behind the libelous articles in *Die Nationale Front* and *Motor und Sport,* and he decided to take legal action against them and Hans Köth, the editor-in-chief of *Die Nationale Front.* After the holidays, in January 1932, he got in touch with his many friends, colleagues, and acquaintances in the auto industry in an effort to collect as much evidence against the three men as possible. One of the first of those prepared to help him was Hermann Klee of Wanderer, who recalled bad experiences with Ehrhardt all too clearly himself. He wrote to Ganz that the "Ehrhardt chapter" at Wanderer was "closed once and for all" and that "no one wanted anything further to do with it." Ehrhardt's letters were left unanswered. Klee went on: "When I think how you spoke up for Herr Ehrhardt and how he owed his appointment by us purely to your recommendation, then I can only sympathize and sense that you too must feel the personal disappointment as deeply painful."[17]

Hermann Klee's contribution was followed a few days later by a letter from Ganz's friend Walter Schneider, an engineer in Cologne. He had known Gustav Röhr since the end of the 1914-18 war and believed him to have had no technical training at all. Schneider wrote that Röhr had been appointed director of the auto manufacturer Priamus by "people who likewise do not enjoy any great renown" and that this had led to the firm's bankruptcy. He added: "In my view Röhr's single undisputed talent is for finding financiers to back the most woolly of plans."[18] Another auto-engineering friend, W. Stoekicht of Munich, had nothing good to say about Röhr, having become caught up in a court case against him over fees for manufacturing rights, and he described his experiences to Ganz in a letter: "Röhr's attempts to influence the case in his favor by means of personal disparagement was clear confirmation of the inferiority as a person that you too have perceived in him."[19]

For the forthcoming court case, along with these witness statements about Paul Ehrhardt and Gustav Röhr, Ganz would need statements in his support to help refute the allegations made against him in *Die Nationale Front* and *Motor und Sport*. Among those he approached were Hans Nibel and Willy Bendit, with whom he had worked closely over the past year and a half. "My opponents are now using the most spiteful means of maligning me," Ganz wrote in a letter to Hans Nibel in which he shed further light on the accusation that his appointment as technical consultant to Daimler-Benz had been made under pressure. "Since these weapons are being deployed in proceedings that do not involve any summoning or questioning of witnesses," he wrote, he would like to ask Nibel to counter them in a formal statement.[20] Hans Nibel replied: "Naturally I am most willing to give you the formal declaration you have requested and to help you in any way I can in your fight against your opponents." In his enclosed statement, Nibel wrote: "The undersigned affirms that at the time of the disagreements between Daimler-Benz and *Motor-Kritik*, and in the settlement that flowed from them, no offer of a position at Daimler-Benz was ever made to Herr Ganz."[21]

Willy Bendit, for his part, declared that over a period of some ten years he had known Josef Ganz to be an "absolutely unselfish man." Although he had initially contested the "methods" of *Motor-Kritik* to some degree himself, he believed Ganz's only purpose lay in "a perfectly natural attempt to promote his magazine" and that he had done so "completely disinterestedly" and purely "in the interests of his readers."[22] Baron Klaus-Detlev von Oertzen, a member of the board at Wanderer, wrote that he had come to know Josef Ganz as an "upstanding, honorable

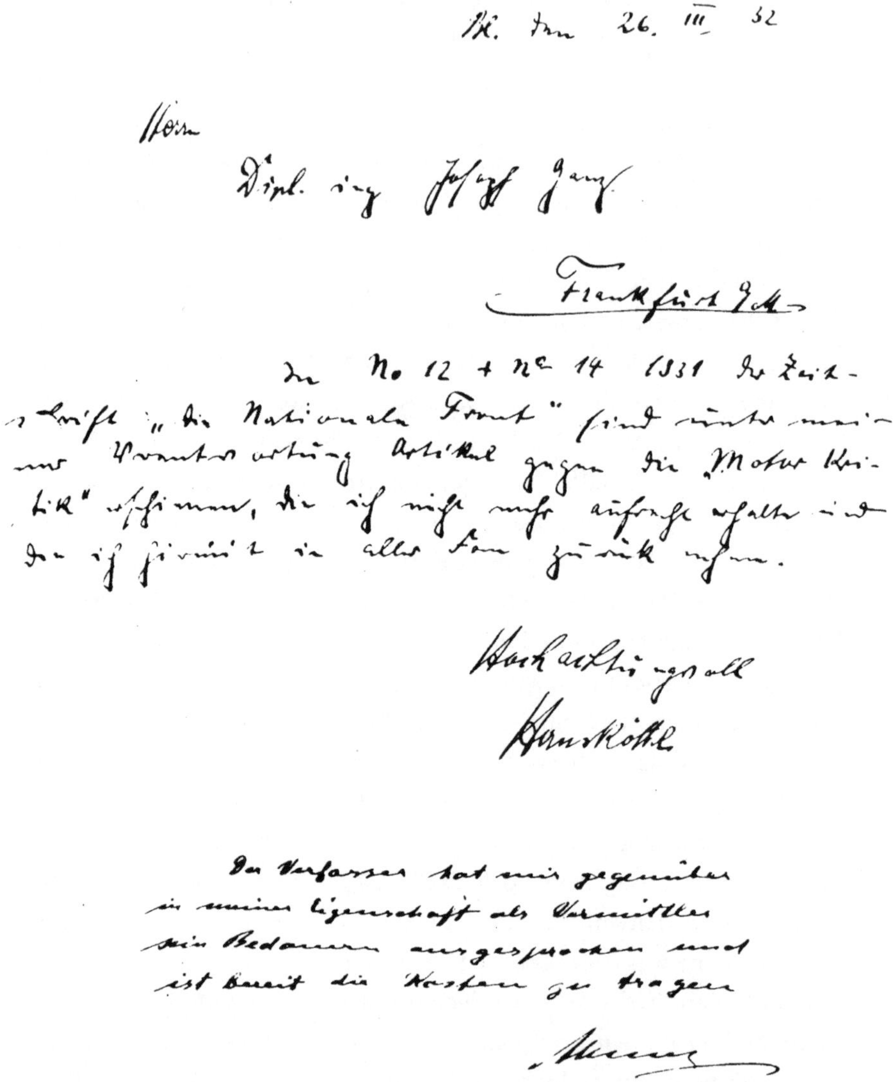

W. den 26. III. 32

Herrn

Dipl. ing Josef Ganz

Frankfurt a/Main

In No. 12 + No. 14 1931 der Zeitschrift „Die Nationale Front" sind unter meiner Verantwortung Artikel gegen die „Motor Kritik" erschienen, die ich nicht mehr aufrecht erhalten und die ich hiermit in allen Form zurücknehme.

Hochachtungsvoll

Hans Köth

Der Verfasser hat mir gegenüber in meiner Eigenschaft als Vermittler sein Bedauern ausgesprochen und ist bereit die Kosten zu tragen

Allmers

Letter by Hans Köth, with comment by Robert Allmers, 1932

person" and as someone possessing "abundant professional expertise." Although his way of working "often did not coincide" with Von Oertzen's own views, Herr Ganz's criticism had arisen from the "purest of motives." Ganz was inspired by a desire to use his journalism to "propel the automobile industry in what he regards as the right direction."[23]

Intervention by the RDA

By late March, Josef Ganz had received a considerable number of supportive statements and was ready to begin his court case, but mutual friends of Ganz and the National Socialist-leaning Robert Allmers, the director of the RDA, purportedly acting "in the interests of the industry," intervened and mediated in talks between the two. Their aim was to resolve the conflict between Josef Ganz and his adversaries. Ganz, who in the past had quite regularly engaged in talks with the RDA, felt that Allmers, "whose position in between the millstones of opposing interests is extremely difficult,"[24] was genuinely doing his best to mediate in the libel case. Allmers urged Ganz not to raise the stakes too high and warned that if he did so he could expect "devastating reprisals" against him personally and against *Motor-Kritik*.[25]

Under pressure from Allmers, Josef Ganz eventually settled for a statement by the editor-in-chief of *Die Nationale Front* Hans Köth withdrawing the allegations. In an untidily scrawled note, Köth wrote: "In no. 12 and no. 14 / 1931 of the magazine *Die Nationale Front*, under my editorial responsibility, articles hostile to *Motor-Kritik* appeared that I no longer stand by and hereby formally withdraw." Beneath Köth's statement, Allmers added: "The writer has expressed his regret to me in my capacity as mediator and is prepared to bear the costs."[26] The whole affair led to technical correspondent Friedländer of *Motor und Sport* being sacked on the spot a few weeks later.[27] The feud between Ganz and Paul Ehrhardt and Gustav Röhr was far from over, however. Indeed, it would grow increasingly vicious.

CHAPTER EIGHT

THE MAY BUG BECOMES THE SUPERIOR

Spring had come to Germany, and Josef Ganz took advantage of the mild weather to make further test runs in his Maikäfer prototype. Not that the cold and snow had prevented him over the past few months. Dressed in his thick leather jacket and gloves, he had taken to the road as usual to see how the little open-topped car would behave on snow and ice. He drove flat out across open country and had recently gotten bogged down in a plowed field covered in melting snow, but even then the Maikäfer had managed to struggle out under its own steam—proving the effectiveness of the concept. The May Bug was so trailblazingly revolutionary that the German auto industry came increasingly under the spell of the idea, as promoted by Ganz in *Motor-Kritik*, fascinated by its independent suspension, swing axles, backbone chassis, low center of gravity, and rear-mounted engine.

The general public too was introduced to the revolutionary design of the May Bug. Josef Ganz was visited in Frankfurt by a film crew from the German newsreel Deulig-Tonwoche. Using a large hand-cranked camera mounted on a heavy wooden tripod, the cameraman filmed him as he demonstrated the car's extraordinary road-holding abilities. He blazed at full speed over a stretch of pavement, drove underneath the trailer of a truck, and accelerated out of a grassy ditch by the roadside. The car shook wildly, its front left wheel coming off the ground, but it took the severe punishment without any problems. The result was a fascinating piece of film, shown to an audience of millions in the German cinemas in the spring of 1932.

Following the success of the Mercedes-Benz 170, Ganz was now involved in the development of a new model with independent suspension at BMW. Along with a BMW test driver, a cheery fellow in knee breeches, a black leather jacket, and a working man's cap, Josef Ganz took a number of test drives that spring in the countryside around Munich with the first prototype of the BMW Automobilkonstruktion München 1 (AM1). Beneath its conventional bodywork, this astonishingly small car, which somehow had room for four adult passengers, was technically highly innovative. Heeding the stern criticism of the previous few months of his dual role as journalist and technical consultant to the industry, Josef Ganz published a fairly restrained piece in *Motor-Kritik* about the BMW AM1 shortly after its official introduction in April 1932. He wrote that as an advisor at both Daimler-Benz and BMW he "did not feel called upon" to decide how well the BMW engineers had designed the car. "Readers of *Motor-Kritik* can do that better themselves, when they test the new BMW." As with the testing of the Mercedes-Benz 170, this would also put readers in a position to decide whether the "design trends championed for many years" by *Motor-Kritik*, such as a low center of gravity and independent suspension combined with a rigid chassis, really did improve driving quality as much as the magazine had always claimed.[1]

Volkswagen "Ideal"

Béla Barényí, an engineer who had become a friend of Josef Ganz, later wrote in *Motor-Kritik* that with the BMW AM1 "the striven for and attainable Volkswagen 'ideal'" had indisputably come very close. BMW, he said, had an opportunity to overwhelm the auto industry completely with this new model, "if only it had the engine at the back!!!!!" Barényí

Jan. 29, 1935. J. GANZ **1,989,446**

DRIVING BLOCK ARRANGEMENT FOR MOTOR VEHICLES

Filed May 2, 1933

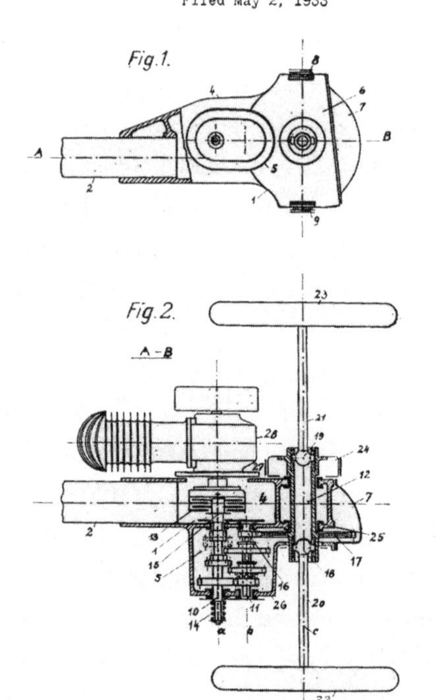

Drawing from US patent #1,989,446,
first filed by Josef Ganz in Germany, 1932

wrote that no one needed "prophetic gifts" to see that such a model would have "a great future." "Bravo BMW!" he declared.[2] BMW, however, did not respond to Barényí's enthusiastic appeal but opted like most other manufacturers for a more cautious and phased introduction of the new construction principles. At this stage, only Daimler-Benz and Zündapp were developing models based on those principles, along the lines of the Ardie-Ganz and Adler Maikäfer prototypes.

Months after the test drive with the Maikäfer at Daimler-Benz, Josef Ganz was still negotiating with technical director Hans Nibel about the possibility of having his design put into production. Nibel wrote him in

June 1932 that "the question of whether a small car can be built according to your ideas depends crucially, first of all, on whether it's possible to produce a car of the size and design you have in mind for a price of RM 1,300." To answer this question, Nibel wanted a full preproduction cost analysis drawn up, and he asked Ganz to deliver detailed drawings for this purpose.[3] Ganz sent Nibel the requested drawings of the structure and assembly of his Volkswagen design, which he had further perfected that spring. The most important innovation was the altered drive mechanism. The engine block was now mounted horizontally on the right-hand side of the central tube of the backbone chassis. To the left was the gearbox, along with the clutch. This gave the car an extremely low center of gravity, right in the middle, for optimal handling. It was an ingenious and progressive design, and Ganz had put in a patent application in May 1932 for the new engine mounting[4] as well as for an improved swing-axle design[5] and a new air-intake system for the engine, whereby the air was drawn in through a grid at the back of the roof, where it sloped down toward the rear of the car.[6] He patented these and other inventions not only in Germany but in Britain, France, Hungary, Italy, Switzerland, Austria, Czechoslovakia, and even Japan and the United States. In Germany, he also applied for two "utility models" (a short-term and less stringent version of the patent), one for the gear-shift mechanism[7] and one for the system by which the electric cables were secured.[8] He now had a comprehensive set of technical drawings and the necessary patent applications were being processed. All he needed was a suitable manufacturer.

Production Plans

That summer, in a brand new Tatra 57, Josef Ganz and Madeleine Paqué went to watch the Herzfahrt, a car rally organized by the German motoring organization, the Allgemeiner Deutscher Automobil-Club (ADAC). The Tatra importer had loaned Ganz the only Tatra 57 in Germany for a full week so that he could write a test report for *Motor-Kritik*. He was extremely enthusiastic and wrote that the car was very close to being the "ideal travel and general-purpose car" as far as power, comfort, and economy were concerned. Its main demerit lay in the new air-cooled, four-cylinder boxer engine, which he found "very loud and turbulent."[9] In the afternoon, when the race was over and he and Madeleine were driving off, Ganz looked in the rear-view mirror and saw a large convertible approaching at speed, flashing its lights to signal him to stop. At the wheel sat *Motor-Kritik*'s editor Frank Martin and next to him a middle-aged man with a moustache, his dark hair blown on end. Ganz parked at the side of the road and greeted Martin, who had not yet seen the new Tatra 57 and was eager to look under the hood and examine the car from all sides. He even lay down underneath it.

Meanwhile the passenger, whom Ganz did not know, got out of the convertible. He introduced himself as Wilhelm Gutbrod, the director of the motorcycle factory Standard Fahrzeugfabrik in Ludwigsburg. Martin had told him about the revolutionary Maikäfer, and he was extremely interested in putting such a model into production. Shortly after that first meeting, Gutbrod came to Frankfurt at Ganz's invitation and was given a full demonstration of the May Bug. Gutbrod said he felt very positive about producing the car, although first he would have to discuss the project with his financial director, Gustav Rau. In July 1932, a manufacturing rights agreement was signed. A Volkswagen would be developed and produced by Standard according to Ganz's patents, and Ganz would

Zu der Patentschrift **576741**
Kl. **46c⁴** Gr. 1

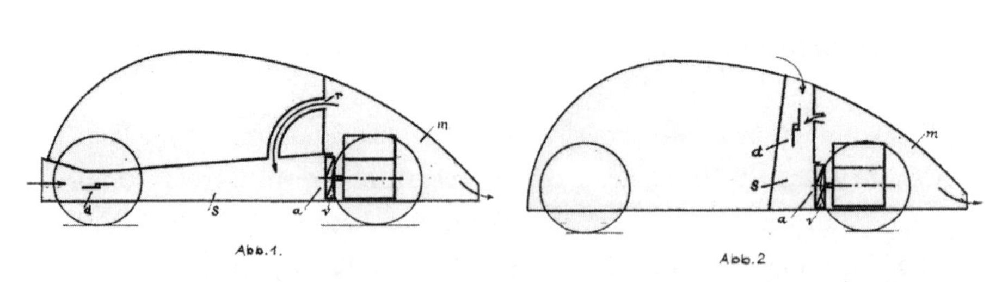

Drawing from DE patent #576741 for a new air-intake system for rear-mounted engines, first filed by Josef Ganz in Germany, 1932

be paid fifteen Reichsmark per car produced.[10] Before the month was out, Ganz dispatched an engineer he employed to Standard in Ludwigsburg with his design drawings. To his great irritation, however, Gutbrod soon started sending him requests for alterations. He wanted closed bodywork and a heavier, two-cylinder engine with an electric starting motor—all of which would increase the weight of the car and therefore its retail price. Ganz warned Gutbrod "not to fall into the errors of all his predecessors" by trying to build a "small Maybach," a reference to one of the most luxurious of German makes of car. His fellow editor Walter Ostwald also advised Gutbrod in no uncertain terms to keep the car as spartan and cheap as possible.[11]

Test Drive across Germany

Like the Maikäfer, the model produced by Standard had a backbone chassis with a central steel tube to the front of which were attached the leaf springs for the independently suspended front wheels, the steering mechanism, and the pedals. At the rear, as described in Ganz's patent, were the horizontally mounted two-cylinder engine on one side of the chassis tube and on the other the gearbox, with directly behind them the swing axles and the leaf springs for the independently suspended rear wheels. This first prototype had simple, open, two-seater bodywork made out of plywood covered in artificial leather, with steel fenders, a floor of corrugated metal, and a canvas sunroof. The headlamps were mounted high up, close to the windshield. Under the hinged front of the car was a small space for luggage and tools, and just ahead of the rear wheels were two large air intakes for the engine. Along with a test driver from Standard, Ganz took extensive test runs through southern Germany that summer, driving the prototype thousands of miles over all kinds of terrain, including nearly forty miles off-road. At every village they drove through, the car attracted a great deal of attention with its rear-mounted engine and a roofline two heads lower than other cars.

The test drives went almost perfectly. Only two cheap, off-the-shelf parts gave out and were replaced in the factory with sturdier alternatives of a different brand. No further alterations were made to the technical design, although Wilhelm Gutbrod decided to develop a different type of closed bodywork for the production model. Josef Ganz tried to convince him that he should start as he meant to go on and use "proper streamlining" as patented by Paul Jaray, but to Ganz's displeasure Gutbrod declined. Instead, the closed version of the car was given a simplified intermediate shape, with a semi-streamlined profile that made it look roughly like a beetle. Since the Standard Fahrzeugfabrik did not have all the facilities available to a car manufacturer, the simple bodywork was built, as in the prototype, out of plywood covered in artificial leather with semicircular metal fenders attached. The sides of the car were straight, with large doors that opened backwards (known as "suicide doors") and decorative trim running from the front to the rear.

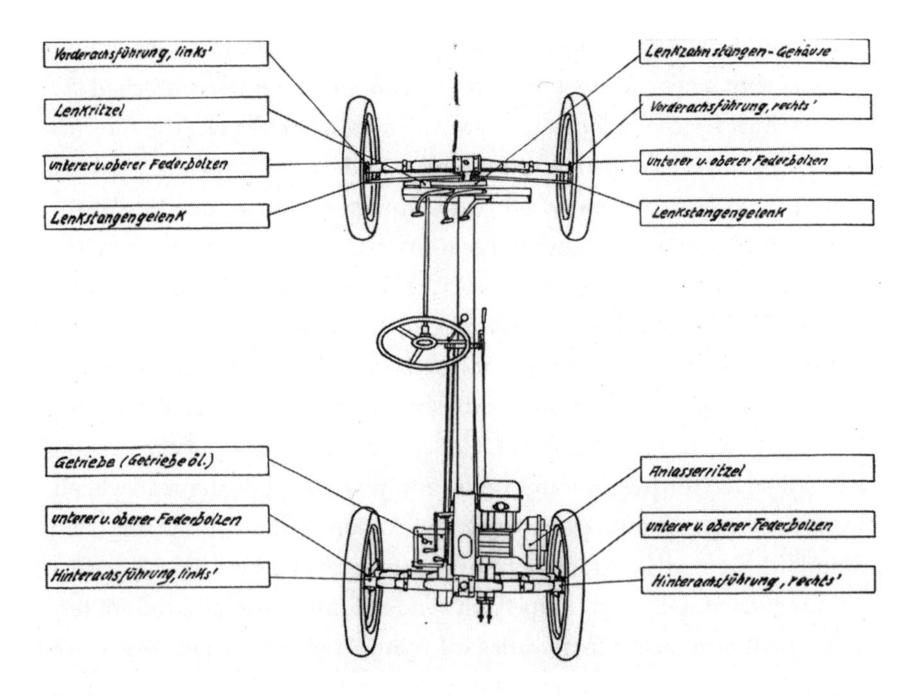

Chassis drawing of the Standard Superior, taken from the German instruction manual, 1933

Ahead of the rounded nose were two headlamps that stuck up like frogs' eyes, and at the lowest point of the tapered roof sat a single rear light, combined with the license plate. Air to cool the engine was drawn in from the back of the roof by the system Ganz had patented. Out of the front of the roof stuck a tiny windshield wiper, and on either side of the front windshield were trafficators, small lighted arrows that hinged outward. The four wheels had simple closed metal rims and thin motorcycle tires. Inside were two front seats and behind them, above the engine, a foot-deep luggage compartment that could also serve as a seat if necessary. The car was only 3,050 millimeters long (a little over ten feet) with a wheelbase of 1,950 millimeters (just under six feet, five inches). The new model was named the Standard Superior, and it would be officially introduced in February 1933 at the International Automobile and Motorcycle Exhibition (IAMA)—the first motor show to be held in Berlin for two years.

The Beetle-like Standard Superior, 1933

Porsche Tests Prototypes

That summer, Josef Ganz and Madeleine Paqué traveled to Switzerland in their faithful old Tatra 11 to visit Uncle Alfred Ganz and his wife Valerie at the Villa Solina. From there they drove on to Czechoslovakia, to the Tatra factory in Kopřivnice, where Josef had a meeting with technical director Hans Ledwinka, a man he greatly admired. During their conversation, Ganz told Ledwinka about the development plans at Standard and confidentially let him see a few of the technical drawings. Ledwinka was intrigued.[12] Although Tatra had been building passenger cars with a backbone chassis and swing axles for years, it did not yet have a model like this in development, with a rear-mounted engine. Meanwhile, at Zündapp the first prototypes of the Type 12 were being tested, built according to plans developed by Ferdinand Porsche's design bureau. The backbone of the chassis had a rectangular profile, and the new model featured independent suspension with swinging rear half-axles. The gearbox was mounted just in front of the swing axles and the engine just behind them. Contrary to Zündapp's wishes, Porsche had given the car a star-shaped five-cylinder radial engine, for which he had also applied for a patent.[13] For the bodywork, a conventional shape was tried first, then a more streamlined version.

According to the initial report on the Zündapp prototype in *Motor-Kritik*, "the great Austrian designer" had developed a vehicle that was "in maintenance costs a small car but in its driving characteristics a comfortable and convenient general-purpose vehicle." The total weight of the chassis with wheels attached was a mere 326 kilograms (719 pounds). "Considering the following data, that's a record: 4 seats, 1,200 millimeter track width, 2,500 millimeter wheelbase, 1.2 liter cylinder capacity." The Zündapp Type 12, however, was not a Volkswagen but a mid-class model, and the report claimed that because the innovative Maikäfer approach had been taken it would be "about 20% lighter—and a little cheaper than current cars of the same size."[14] The Type 12 never came onto the market; Zündapp put the project on hold after encountering problems with the road holding; the engine, which proved too noisy; and the high production costs entailed by the self-supporting steel bodywork. In any case, there had been a sudden and marked increase in demand for new motorcycles, so Zündapp could not free up enough production capacity either for a Volkswagen or for its planned new racing car. Ferdinand Porsche was allowed to keep one of the three Type 12 prototypes for private use. He transferred the racing car project, for which Zündapp had already built an impressive sixteen-cylinder engine,[15] to the newly established Auto Union, made up of DKW, Wanderer, Horch, and Audi.

The Modern Kommissbrot

Abroad too there was growing interest in the Volkswagen concept. In 1932, the Czechoslovakian auto manufacturer Skoda built a prototype it called the Type 932, with a backbone chassis, independent suspension, and a rear-mounted, air-cooled, four-cylinder boxer engine. It had four seats and a snub nose reminiscent of the Mercedes-Benz 120H prototype. As had happened at Zündapp, however, Skoda decided against putting the car into production.

Yet another manufacturer that tried to build a Volkswagen based on the Maikäfer design in 1932 was Hanomag. Four years after ceasing production of the Hanomag 2/10 PS Kommissbrot, the company had commissioned the German engineer Pöllich to design a "modern Kommissbrot," but because of the credit crunch the company found itself in severe financial difficulties and was forced to suspend development of the prototype.[16] In a desperate attempt to discourage the competition, Hanomag issued a "sensational" press release in August 1932, revealing the existence of its secret prototype. At *Motor-Kritik,* the editors read that:

> From time to time it is announced, almost dictatorially, that only front-wheel drive, or only rear-wheel drive, or only swing axles ought to be indispensable structural characteristics for a motor vehicle. [...] Today this idea (and all technicians in the world should perk up their ears at this point) has been definitively abandoned! A design is relinquished, years of brainwork are presented to the world, and development is broken off as a blind alley! If only all auto manufacturers were to report not just on their successes but on their failures as well! We'd already be a good deal further along the road and able to avoid much fruitless double work!

In the same press release, Hanomag revealed "the new Hanomag, the modern Kommissbrot, the Volkswagen," which would "fulfill all the user's wishes," yet "nevertheless: it's not our Volkswagen!"[17] The company went on to explain the innumerable things that were wrong with this way of building cars—all nonsense according to Josef Ganz, who responded in *Motor-Kritik* that it was "horrifying to see "the trivialities" over which "the Hanomag engineers have stumbled," especially "when you consider how little creative power has been applied to tackling the greatest problem in car building." All things considered, he was not sorry development had been halted. "We would advise a return to the order of the day, after taking due note of all this." He told his readers that fortunately developments right across the industry were so advanced already that "no power any longer has the ability to delay the advances that will at last give us the perfect automobile."[18]

Ganz was right. The wave of innovation that had already produced radical models like the Mercedes-Benz 170, the BMW AM1, and the front-wheel-driven Adler Trumpf was now unstoppable. "I would modestly maintain that *Motor-Kritik* has made an outstanding contribution to this new technical orientation," wrote Béla Barényí that summer in the magazine, and he posed a question to readers:

> Haven't people almost overexerted themselves already in complimenting Ledwinka, Porsche, and Röhr? Were the written treatments by Ganz, Janus, Arnau, O Weller [...] and others not all perfectly accurate sketches of a time in which the battle with technical and economic fictions seemed to have been fought in vain?[19]

The argument that had been raging for years over the value of certain technical innovations had turned into a neck-and-neck race to see who could be the first to put a modern car incorporating Maikäfer construction principles on the market. After the halting of developments at Zündapp and Hanomag, Standard had stolen a march on the others.

Tatra Versus Ganz

In the fall of 1932, with an eye to the forthcoming introduction of the Standard Superior, Josef Ganz carried out extensive research into published patents relating to the use of a backbone chassis. While leafing through paperwork in the library at the Patent Office, his eye fell on a peculiar transaction. To his amazement, in August Tatra appeared to have bought two eight-year-old patents from his archrival Paul Ehrhardt, who had himself originally purchased the first of them[20] in 1925 from the German inventor Arnold Seidel. He had applied for the second[21] that same year as a supplement to the Seidel patent. Both described a rather remarkable design for a car with a backbone chassis and swinging rear half-axles. Ganz expressed his astonishment in *Motor-Kritik*: "Hardly anyone is going to build such a thing, are they?" He went on: "In which case it's surprising that the Tatra factory, holder of the patent, is nevertheless paying the considerable costs of regularly renewing it. What do they have in mind?"[22]

Tatra's intentions quickly became clear when Ehrhardt submitted an official complaint against BMW in writing, saying that two of the patents used in the BMW AM1[23] amounted to theft of his intellectual property. At the request of BMW, technical consultant Josef Ganz studied the patents and the complaint. He advised BMW to take legal steps against Ehrhardt. Ganz was convinced BMW had not violated Ehrhardt's two patents and that the case had been brought purely to cause disruption in the industry, to hinder progress, and as an unjustifiable attempt to acquire "ransom money."[24] Ganz suspected that Tatra was trying to shore up its competitive position through coercion, now that more and more cars with swing axles were appearing on the market and the company was in danger of losing its lead. In late 1931, immediately after the introduction of the Mercedes-Benz 170, Tatra had developed a steering mechanism with independent suspension in the front wheels that was practically identical to the system used in the Mercedes.[25] The conflict with BMW later resulted in a court case in Frankfurt, in which Ehrhardt's claims were indeed dismissed.[26] This meant in practice that the two patents Tatra had bought from Paul Ehrhardt were worthless. Concurrently with the case against BMW, however, Tatra decided, with Ehrhardt's support, to apply for a series of new patents with the aim of thwarting the patent applications made by Josef Ganz, which were still pending. Ehrhardt had undertaken a similar attempt at fraud less than ten years earlier when, at Luftschiffbau Zeppelin, he had tried unsuccessfully to lay claim to Paul Jaray's streamlining patents.[27]

In late 1932, Tatra put in patent applications for a swing axle design,[28] for the mounting of swing axles on a backbone chassis,[29] and for the mounting of an engine block on a backbone chassis.[30] Ganz studied the various applications and in the case of the swing axle design he immediately noticed an extremely close resemblance to the rear axles he had used on the Ardie-Ganz prototype. The third patent, for the mounting of an engine block on a backbone chassis, was identical to a design that the engineer Arnold Seidel had described in *Motor-Kritik* two and a half months earlier.[31] During a working visit to Daimler-Benz in Stuttgart, Josef Ganz discussed all this with Wilhelm Kissel and Hans Nibel, since Daimler-Benz also wanted to produce a car with a backbone chassis, swing axles, and a rear-mounted engine. If Tatra's patent applications were successful, Ganz argued, then the entire auto industry would soon be obliged to pay royalties to Tatra on all cars of this type that were built. Kissel and Nibel promised to help him block the Tatra patent applications, and he found another ally in the auto manufacturer NAG (Nationale Automobil-Gesellschaft). Josef Ganz, Daimler-Benz, and NAG signed a joint protest against Tatra and submitted it to the Patent Office in Berlin. Tatra would succeed in its underhanded ploy only if it could prove that its designs predated Ganz's work on the same innovations. To this end, it produced in court a photograph of a prototype and a technical drawing of the swing axle design it claimed to have developed. When Ganz saw the material he was amazed to find himself looking at a stolen photograph and drawing of his own Ardie-Ganz prototype.

He approached Ardie, where the Ardie-Ganz had been built in 1930, and was told that neither the drawing nor the photographic material were any longer in its possession. No one knew where they had gone, but Ganz suspected Ehrhardt had gotten hold of both the prototype and its documentation and simply made off with them.[32] Ganz was fortunate enough to have kept the original negative of a photograph showing him standing at his drawing table working on that same drawing, as well as photos of the drawing itself and of the Ardie-Ganz prototype. Moreover, the picture of him at his drawing table, taken by Madeleine Paqué in 1930, had appeared on the front cover of the first January issue of *Motor-Kritik* in 1931. During this battle over patents, *Motor und Sport* published a critical article in its December 1932 issue about so-called patent pirates, in which the anonymous author pointed out that a journalist was obliged to serve the interests of the public and a structural engineer the interests of his company. It was therefore no more the task of a structural engineer to apply for patents that hindered his company's development than it was the task of a journalist to work in such a way that the interests of a particular manufacturer were served.[33] From the tone of the article, Josef Ganz deduced that this was a direct criticism of him by Ehrhardt, and he responded with a short piece in *Motor-Kritik*, observing that "behind the anonymous author lurks a patent hunter who is using the magazine to popularize opinions that suit his purpose."[34]

Josef Ganz was determined at all costs to prevent Tatra and Ehrhardt from fomenting disquiet in the German auto industry and thereby standing in the way of developments. By discussing in *Motor-Kritik* all existing patents relating to, among other things, the backbone chassis, swing axles, and steering mechanisms, Ganz was attempting to bring out into the open any overlapping of patents. To reinforce their ties even further, Daimler-Benz entered into a second contract with Ganz in January 1933 that obliged him to offer all his future inventions first to Daimler-Benz. In return, he would be paid RM 250 per month on top of his salary of RM 500 as technical consultant.[35] Under the new agreement, Ganz would need to ask official permission from Daimler-Benz before he could introduce the Standard Superior at the international motor show, the IAMA, in Berlin that coming February. On the eve of the show, Hans Nibel replied that Daimler-Benz would not make use of its "contractually established right to veto the granting of a license to the Standard factory." As a condition, he insisted that "no further essential changes" should be made and that on no account should the finished vehicle weigh more than five hundred kilograms (1,102 pounds) or have a more powerful engine.[36] Nibel was trying to ensure that the Standard Superior would not become a rival to a new mid-class model with independent suspension, a backbone chassis, and a rear-mounted engine that Daimler-Benz, with Ganz's help, wanted to put on the market within the coming year.

CHAPTER NINE

PRESENTATION FOR HITLER

On an icy winter's morning at the start of February 1933, Josef Ganz left early for Frankfurt's main station. At the kiosk in the station concourse, he bought a selection of newspapers and magazines for the long train journey to the international motor show, the IAMA, in Berlin. The papers were full of news about Adolf Hitler, who had been sworn in as chancellor by President Paul von Hindenburg less than two weeks before, heading a coalition government in which his own Nationalsozialistische Deutsche Arbeiterpartei (the NSDAP or Nazi Party) was allied to the Deutschnationale Volkspartei (DNVP). Ganz had his suspicions about the anti-Semitic new chancellor, but at the same time he was impressed by the decisiveness displayed by the new cabinet. Immediately on coming to power it had declared an intention to provide active support for the motorization of Germany. As a first step, the Ministry of Transportation announced in early February that as well as three-wheelers, small four-wheeled cars with a maximum engine capacity of 0.4 liters could be driven by holders of a motorcycle license. For Josef Ganz, this was a godsend. The Standard Fahrzeugfabrik in Berlin was about to introduce the Standard Superior, based on his patents, a small four-wheeled Volkswagen that would be marketed in two versions, with an engine capacity of either 0.4 or 0.5 liters.

This change in the law—for which Josef Ganz had been campaigning for five years—would open the way for the further development and widespread distribution of the fully-fledged German Volkswagen. In his train compartment, Ganz leafed through the latest issue of *Motor-Kritik* one more time. It had been published just days before the start of the IAMA. He was so pleased with the advances in the German auto industry that he had written to readers in his opening editorial:

> *Motor-Kritik* is proud to have been involved with the latest developments, which have put Germany back into a leading position in car building. And it has a sense that because of its success across the board the magazine has more than safeguarded itself from everything in the way of incomprehension, legal action, and abuse it has had to put up with over the years of your struggle and work for that goal, now almost attained. What we have achieved first and foremost is to have overcome stagnation in the design of automobiles—it is even safe to say conclusively overcome. The vast majority of factories have penetrated virgin territory so deeply that for them there is simply no way back. That is a fortunate thing—for the future belongs to those who think ahead.[1]

Zündapp in the Snow

Josef Ganz felt a thrill of excitement when the large black express train from Frankfurt steamed into Berlin's main station several hours later. Through the clouds of smoke and steam he made his way to the station concourse and out across bustling Berlin to his hotel. The next morning, he took the subway, the most reliable mode of transportation in the heavy February snow, to the exhibition complex on the Kaiserdamm, where later that day the brand-new German chancellor would officially open the International Automobile and Motorcycle Show, the IAMA. In the early morning, there was a hectic coming and going, with trucks still delivering exhibition models. At the very last moment, an open truck from Ludwigsburg arrived, bearing the Standard Superior and an additional chassis in a large wooden crate.

Ganz had the honor of driving the car to the entrance, and he felt unable to resist the opportunity to run it up and down the snow-lined Kaiserdamm for a few minutes. Through the little semicircular front windshield, cleared of snow by its tiny windshield wiper, he suddenly saw one of the Zündapp Type 12 prototypes at the side of the road. He braked and stopped next to Ferdinand Porsche's private car, grabbed his camera, and took three photographs of the big brother and little brother together, two advanced models based on his Maikäfer design. His colleague Walter Ostwald had recently taken a test drive in the Zündapp Type 12 and concluded in *Motor-Kritik*: "In this prototype answers have been found to a lot of acute problems." He believed it was a fundamentally progressive design. The car exhibited excellent road holding and driving comfort, and it had roomy streamlined bodywork that represented a good compromise between "aerodynamic resistance and the resistance of public taste." Ostwald had discovered only a few minor drawbacks—a noisy exhaust, a stiff gear-change, and an overly prompt response from the gas pedal—all of which would be easy to fix for serial production. He therefore hoped the car would quickly be made available to the "clientele that is undoubtedly on hand," since "we need an economical and properly functioning Volkswagen."[2] But development of the Zündapp Type 12 had already been halted. After examining the car, Ganz stepped back into the Standard Superior, the first production model built along Maikäfer lines, and drove to the entrance of the IAMA, leaving Porsche's Zündapp behind him in the snow. The parked car was a sure sign that Ferdinand Porsche was already on the exhibition floor, walking around among many other engineers and industrialists, including Hans Nibel, Wilhelm Kissel, Hans Ledwinka, Gustav Röhr, and Joseph Dauben.

Hitler's Speech

While the exhibitors were putting the finishing touches to their stands, Josef Ganz saw Hermann Göring, Hitler's right-hand man, pacing the exhibition floor as he waited for the official opening. Göring stopped abruptly at the Standard exhibit. Flanked by motorcycles, the chassis of the Standard Superior stood diagonally across one corner of the stand. The complete car, with its simple beetle shape, stood next to it. Göring seemed greatly interested in the little Volkswagen, but he hardly had time to look.[3] Near the stage the dignitaries were gathering, including Chancellor Adolf Hitler, Propaganda Minister Joseph Goebbels, the director of the Reichsverband der Automobilindustrie (RDA) Robert Allmers, and Adolf Hühnlein, the leader of the National Socialist Motor Corps (the Nationalsozia listisches Kraftfahrkorps or NSKK, an automobile association that also taught driving skills and was organized along paramilitary lines). Allmers was the first to speak. He had obstructed Josef Ganz and his *Motor-Kritik* significantly in recent years, but he could no longer ignore the huge wave of innovation in the auto industry for which the magazine had consistently campaigned. In his speech, he claimed that low import taxes on foreign cars had been responsible for the difficulties faced by the Germany car industry over the previous few years. A cynical-looking Josef Ganz watched and listened. It was precisely because of foreign competition, he thought to himself, that the German industry had at last begun to innovate. High import tariffs were "poison with a conservative effect."

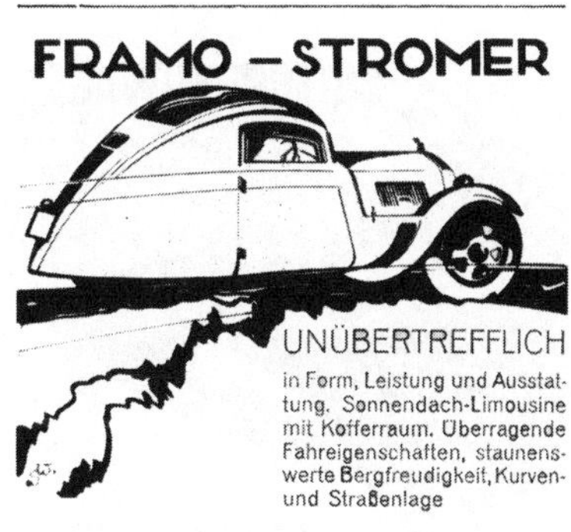

Ad for the three-wheeled Framo Stromer, 1933

Ganz was rather more impressed by the speech delivered by Hitler immediately after Allmers had finished. Hitler recognized and underlined the great importance of road transportation and of the private car in particular. He was determined to support and stimulate the further development of auto engineering by reducing the tax burden on drivers. He also favored financial support for motor racing as a means of boosting technological developments. He ended his opening speech with hearty expressions of thanks to the industrialists and the brilliant designers and technicians for their efforts over the past year.[4] Ganz felt that Hitler's expression of gratitude was personally addressed to him. After all, the February 1933 exhibition was a great showcase for *Motor-Kritik*; several German manufacturers had developed new models over the past couple of years that used the construction principles long advocated by his magazine, such as independent suspension, swinging rear half-axles, a lightweight backbone chassis, and the first signs of streamlining. Conservative car builders were steadily losing support, and this IAMA quickly became known as "the show of shows."[5] After the opening speech, Hitler and his entourage took an extensive tour of the exhibition floor, followed at a respectful distance by an army of press photographers. Among them was editor-in-chief Josef Ganz, carefully recording all the new developments on view at the IAMA for *Motor-Kritik*. Hitler was a great car enthusiast and displayed quite a bit of technical knowledge, much of which he had learned from his close personal friend Jakob Werlin, a Mercedes-Benz dealer in Munich, who had supported him since the early 1920s. Hitler was particularly drawn to the innovative models from Mercedes-Benz, Röhr, and BMW, and from Maybach, which had chosen this occasion to present its luxury streamlined limousine, the Zeppelin DS8.

Ad for the three-wheeled Hercules, 1933

Slowly the procession moved on to the smaller manufacturers, Standard among them. From a short distance away, Ganz watched as Hitler expressed great interest in the little Standard Superior to the company's director Wilhelm Gutbrod and financial director Gustav Rau.[6] Gutbrod explained to Hitler that the car had been developed according to an entirely new concept, with a backbone chassis, independent suspension, swing axles, and—unlike any other four-wheeler at the show—the engine at the rear, mounted horizontally in front of the rear axle, giving the Superior unparalleled road holding. Gutbrod told Hitler and his entourage that the unique chassis design was based on patents held by Dipl.-Ing. Josef Ganz, editor-in-chief of the progressive specialist magazine *Motor-Kritik* and a champion of the small car. Even with the modest numbers Standard had in mind for serial production, the price of the car would be only RM 1,590, and in mass production it could drop even further, to around RM 1,000, the figure Josef Ganz advocated as the retail price of the "German Volkswagen." With the Standard Superior, Ganz had almost entirely fulfilled his promise of two years before, when he had told visitors to the 1931 Berlin motor show that he would produce a streamlined car with swing axles for only RM 1,500. Hühnlein, charged by the new government with motorizing the German people, exclaimed enthusiastically to his chief of staff: "The first man you must engage to work for us is Dipl.-Ing. Ganz!"[7] He wanted the NSKK to take the Standard Superior for a thorough test drive.

Massacre in Ludwigsburg

The Standard Superior, built by the Standard Fahrzeugfabrik in Ludwigsburg, was a huge hit with the public. After Hitler and his retinue left, the doors were flung open and crowds flooded into the vast exhibition halls. Despite the fabulous luxury vehicles on show from prestigious firms like Maybach, Mercedes-Benz, Austro-Daimler, and Röhr, visitors and journalists from all over the world crowded around the Standard exhibit to catch a glimpse of the Superior, the smallest yet most innovative four-wheeler in the entire exhibition complex. The chassis, designed purely with functionality in mind, was a technical work of art. All day, people thronged around it. Visitors in their long, thick winter coats stopped to see what was so special about the car. Josef Ganz was particularly honored by the presence of Edmund Rumpler, who stood next to his own "brainchild," a car directly inspired by the Tropfenwagen.

Ganz was proud of the overwhelming attention paid to the Standard Superior in Berlin, but as he had already explained at length in the latest issue of *Motor-Kritik*, despite his protests Standard had transformed the "sweet little Maikäfer" into an "almost feudal vehicle." "That was a massacre in Ludwigsburg!" he declared. He wrote that he was not at all in favor of the extended chassis, for example, or the choice of a larger engine, the electric starter motor, bigger wheels, or any number of other detailed changes. They made his heart bleed. He was particularly disappointed by the bodywork, which he described as "hideous and impractical," adding: "either use proper streamlining or a decent standard shape." He did have to admit that the car hugged the road well and had "satisfactory" suspension and steering. Ironically, that same issue of *Motor-Kritik* featured a full-page advertisement for the Standard Superior, with the enthusiastic slogan, "all the desires of MK readers fulfilled."[8] It was the first and last advert by Standard in the magazine. Gutbrod was so piqued by Ganz's comments that from then on he advertised in the rival *Motor und Sport.* He responded to Ganz's critical article with a letter printed in *Motor-Kritik* under the title "The May Bug Comedy," a reference to a popular children's book of the day. "You are without doubt lucky that you found us to make your ideas a reality," Gutbrod wrote. "'Young design engineers' hearts bleed readily and easily. That doesn't matter as long as no billfold bleeds with them. Preventing that was my duty. What you and a few diehards want is by no means what the rest of the public wants."[9]

Reichstag Fire

The general public had thirteen days in which to gaze in admiration at the Standard Superior and the many other new models at the IAMA. A few days after the show was dismantled and the last trucks left to return the hundreds of cars and motorcycles to their German factories, Berlin suddenly became the setting for hectic political developments. On the evening of February 27, 1933, at around ten o'clock, the Berlin fire brigade was alerted to a fire at the Reichstag, the seat of the German parliament.

Sirens blaring, a convoy of shiny red fire trucks raced through the icy streets of Berlin to the landmark building, where flames and enormous clouds of smoke were already climbing high above the large glass dome in the center of the roof. After battling the blaze for an hour and a half, the firemen had it more or less under control. At that point, Hitler and Goebbels drove up and were met by Göring, who had reached the spot suspiciously quickly. Göring shouted: "This is the start of the Communist revolt! They'll launch their attack now! There's not a minute to lose!" Hitler in turn told Vice-Chancellor Franz von Papen that the fire was "a God-given signal," and, if indeed, as he believed, the fire was the work of the Communists, "then we must crush this murderous pest with an iron fist!" In the political witch-hunt that followed, the Nazis indeed blamed the Communists, but among opponents of the new regime there was a strong suspicion that the mysterious fire had in fact been started by the Nazis, possibly even by Göring himself.[10]

In the turbulent days after the Reichstag fire, Josef Ganz, at home in Frankfurt, wrote his initial account of the Berlin motor show for *Motor-Kritik*. How different this show had been from its predecessors. "The exact opposite of monotony," was his first impression of the models he had seen. "At last a general realization has dawned that mass production of imperfect cars, however cleverly thought out, is wrong." He expressed respect for the many little three-wheeled cars on view at the IAMA, including the Framo Stromer, Rollfix Rekord, Hercules, Tamag Zepp, Theis, and "the first and still leading" Goliath Pionier, of which the open-topped model was a "bargain" at RM 1,090. The closed version was on sale at RM 1,390, whereas the four-wheeled Standard Superior cost RM 200 more. Mass production in its modern assembly-line factory had enabled Opel to offer a conventional car with four seats and a one-liter engine at a "sensationally cheap" RM 1,990.

In a photo collage with the caption "Swing axles everywhere!" Josef Ganz featured several of the many makes that now used them, whether in their passenger vehicles, delivery vans, or trucks.[11] Further on in the magazine was a reader's letter in the form of a controversial collage with the title "Behind the Official Back of the Show!" and underneath it a photograph of Hitler cut out of a newspaper and a news clipping that read

"Reichschancellor Adolf Hitler thanks the gentlemen of the industry, the brilliant designers and technicians, and the great army of Germany workers for their unwavering diligence." In the photograph, Hitler looks to his right, at the title of the magazine, which the reader and contributor believed was manifestly where the chancellor's expressions of gratitude should be directed. Below the collage were the words:

> My dear Herr Ganz, should I not make something clear in the speech? We want to be honest, after all. Many people have Ursinus, yourself, the Maikäfer, your staff, in short the steadfast work of the magazine *M-K* to thank for the ideas that gave rise to expressions of gratitude in Berlin in 1933.[12]

In that same motor-show issue was a letter from Oskar Ursinus, the founder of *Klein-Motor-Sport*, the magazine out of which Josef Ganz had created *Motor-Kritik*, who wrote that his visit to Berlin had been an "extraordinary surprise, a bright spot." He went on: "Remember the situation a few years ago! I can only say, terrific progress!" Moreover, "all the dreams of 1920" had been fulfilled now that "a fourth wheel" had been added to the small car. This had finally been achieved with the Standard Superior. Reports began appearing in the press at home and abroad in early March. The first was an article in the official army magazine *Der Kraftzug in Wirtschaft und Heer* about the innovations on show at the IAMA, including an assessment of the Standard Superior, which had recently been subjected to an extremely thorough road test by the NSKK.[13] Dipl.-Ing. Baurat W. E. Fauner said of it:

> Entirely new is the Standard, developed by Ganz. The chassis [...] makes a very neat impression. The little 0.4 or 0.5 liter two-cylinder engine is placed horizontally (lower center of gravity) between the rear wheels. Countless test runs have shown that this chassis possesses excellent driving qualities.

Like Josef Ganz, Fauner was critical of the bodywork of the show model, calling it "nonviable in this form." He wrote: "It looks unattractive and the accessibility of the engine is utterly inadequate." And the water and fuel filler openings had been awkwardly positioned in the "streamline-style rear." Fauner commented that this chassis deserved "a particularly well-considered superstructure."[14] In its survey of the cars on show at the IAMA, *Motor und Sport* made a comparison between the shape of the Standard Superior and that of the old Hanomag 2/10 PS Kommissbrot, and it appealed for a "fundamental refinement of the styling, which will then make this vehicle into an extremely elegant and fast car." The chassis, after all, had a "very interesting structure" that gave it "exceptional road holding." The "relatively smooth design of the whole chassis" would help ensure that "the air-cooling of the rear-mounted engine works perfectly, something that is perhaps especially noteworthy with an eye to further development of the rear-mounted engine."[15] Standard proved receptive to criticism of the rudimentary bodywork it had used on its extremely innovative chassis and immediately, in April 1933, produced a new design. The basic shape remained the same, but the lines were more fluid, and the front windshield was slanted to either side at a shallow angle. The new round nose had a slight upside-down V-shape on its upper side, giving the bottom of the five-sided windshield a characteristic kink. That month, the first of the cars were delivered to dealers, but the extra investment had forced Standard to raise the price from 1,590 to 1,620 Reichsmark.

Crowds around the Superior

Immediately after the introduction of the new bodywork, the engineer Werner Fuess carried out an extensive test drive in the Standard Superior for *Motor und Sport*. He was especially impressed by the chassis, which struck him as "exceptionally simple and clear." Fuess also noticed that the Superior was far roomier inside than the other small cars he had seen at the IAMA in February. The two "suicide doors," at almost a yard wide, made it easy to get in and out. Behind the two spacious front seats was a foot of luggage space across the full width of the car. One thing Fuess did find inconvenient was the absence of a rear windshield in this first model. The engine sprang into life immediately when he pressed the starter button, and Fuess quickly mastered the controls. The long springs of the suspension meant the car could be taken on "unimaginable off-road trips, even over fairly poor terrain," and it exhibited "fabulous maneuverability" that would "certainly be experienced as highly agreeable in traffic." Another important advantage, in Fuess's view, was that because the engine was at the rear, noise was limited, "such as one does not expect as a rule in a small car." Fuess concluded: "The basis of the structural design definitely shows, as do practical tests with the Standard car, that it is possible to solve the problem of the manufacture of modern small cars by adopting entirely new ways of thinking."[16] Despite his initial criticism, Ganz developed an increasing affinity for the little Standard Superior, and he enjoyed driving it. Wherever he parked in the still near-empty streets, people thronged around, just as they had around the Maikäfer. Pedestrians stopped and cyclists dismounted to take a look at it. Some of them had already read about the new "German Volkswagen" that you could drive with a motorcycle license. The Standard Superior's new bodywork was first depicted in *Motor-Kritik* in early April, in a jubilee issue celebrating the thirtieth birthday of the swing axle that opened with the message, in bold type:

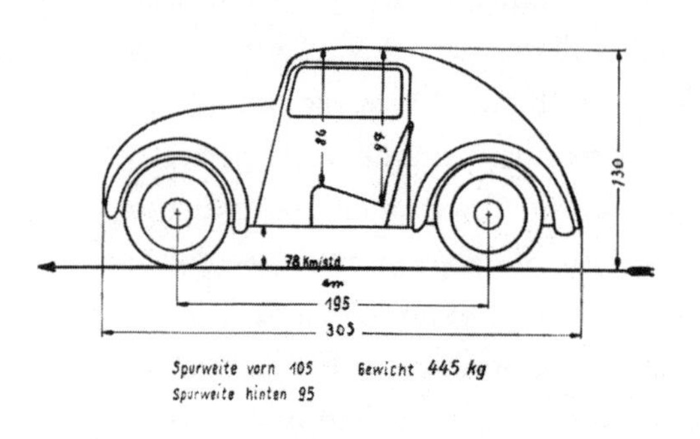

The small dimensions of the Standard Superior, 1933

> The starting shot for the reconstruction of Germany has been fired. After years of aimless wandering, motorization has been recognized as a vital national interest and placed at the forefront of the nation's mobilization. Friends of *Motor-Kritik*, things we have not dared to dream in our wildest fantasies are about to become a reality. The new Germany must be a motorized Germany. Ensure that all available forces are deployed effectively to this end. Join in, so that the aims announced by our government can quickly be converted into action.
>
> —*The publisher and editors*[17]

Aside from the motorization so warmly acclaimed by *Motor-Kritik*, the "new Germany" was becoming a more and more horrific place. Immediately after the mysterious Reichstag fire, Hitler secured a stranglehold on the country by introducing an emergency decree[18] that deprived the German people of a number of basic rights, followed by emergency legislation[19] in late March that concentrated all power in his hands. Many communists and other "enemies of the state" were rounded up and detained in the recently opened Dachau concentration camp. These were the first firm steps toward a totalitarian dictatorship. Anti-Semitism became active government policy in April 1933, with a nationwide boycott of Jewish shops, lawyers, and doctors and the sacking of Jewish government employees by law, with the exception of World War I veterans, those who had been in the civil service since the start of the war, and those who had lost a father or son in combat. A similar law was passed for the sacking of Jewish lawyers. As one way of building support for its reign of terror, the National Socialist government had a considerable interest in bringing Jewish plots to light and in wiping from the collective memory the hugely positive influence of German Jews on German science and learning.

One aspect of this policy was the celebration of "genuinely" German heroes and pioneers. In Mannheim in April 1933, a grand gathering took place to mark the unveiling of a statue of Karl Benz, the German pioneer of the auto industry who had died four years earlier at the age of eighty-four. To the great annoyance of the Ministry of Propaganda, a memorial bust of a Jewish auto pioneer, Siegfried Marcus, had been erected in front of the Technical College in Vienna only six months before, and Marcus' first car, built in 1870 and therefore preceding Benz's by fifteen years, was on prominent display at the Vienna Technical Museum. On that occasion, Josef Ganz had celebrated Marcus at length in *Motor-Kritik* as the inventor of the automobile with an internal combustion engine.[20] It would be several years yet before the Nazis annexed Austria and attempted to destroy the bust, the car, and the rest of Siegfried Marcus's legacy, but in their ceremony that spring day in Mannheim in 1933 they focused international attention on Karl Benz as the true inventor of the motor car and made a firm connection between him and the work of Adolf Hitler.

The gathering in Mannheim was a major event that brought together auto engineers, manufacturers, and journalists from home and abroad. Josef Ganz, a Jewish journalist in Nazi Germany, was still able to take part in events like this without constraint. The day began with the unveiling of a memorial plaque on the house where Karl Benz had lived. His widow, Bertha Benz, answered questions from the press. Then the hundreds of guests moved on to Mannheim Castle, built in 1760, one of the largest Baroque structures in Europe, for a festive lunch in the great hall followed by the unveiling of the statue. Surrounded by red-and-black swastika flags, famous Nazis gave speeches, including Freiherr von Eltz-Rübenach, the transportation minister; Robert Allmers of the RDA; and Adolf Hühnlein of the NSKK.

That afternoon, motorcyclists representing the SA, the paramilitary security organization set up by Hitler in 1921, rode through Mannheim three abreast, followed by a procession of historic passenger vehicles, racing cars, and trucks, from pioneering models by manufacturers including Benz, Daimler, and Opel to modern vehicles from companies like Mercedes-Benz and Maybach. Edmund Rumpler took part at the wheel of his first Tropfenwagen, followed by a Benz Tropfenwagen and other racing cars. A film crew joined the convoy to record the entire spectacle for posterity.[21] Josef Ganz shot several rolls of film, capturing all the unique hardware that drove past him that day and many famous faces from the auto industry. Hans Ledwinka and Baron von Ringhoffer had come all the way from the Ringhoffer-Tatra works in Czechoslovakia. In a top hat and monocle, Ringhoffer smiled at Ganz with a look of rather cynical surprise but did not divulge what he knew to be happening at that very moment at the editorial offices of *Motor-Kritik* in Frankfurt.

CHAPTER TEN

GESTAPO CELLAR ON THE ALEXANDERPLATZ

Josef Ganz was still in Mannheim when, shortly after lunch on a rainy spring afternoon in late April 1933, there came a loud knock at the front door of the stately villa on the Zeppelinallee in Frankfurt. A male voice shouted: "Open up! Gestapo!" The owner of the house, Madeleine's mother Maria Paqué-Köpp, opened the door with trepidation. She found herself facing an agent from the state intelligence service, the Sicherheitsdienst (SD), along with two men from the Geheime Staatspolizei, or Gestapo. The SD agent stepped into the hall in his muddy black leather boots and demanded immediate access to the *Motor-Kritik* editorial office on the floor above. He stamped up the stairs and yanked open the door to the office, where at that moment only Madeleine Paqué and editor Georg Ising were working. Madeleine instantly recognized the SD agent's rather shifty face and slicked-back hair. It was Paul Ehrhardt, the opportunist ex-employee of *Motor-Kritik* who had turned against the magazine and its editor-in-chief Josef Ganz less than two years before. Ehrhardt was not known to Georg Ising, a middle-aged engineer who had only recently joined the magazine after a career in the industry and who lived with his wife in another part of the house. Ehrhardt stood in the doorway for a few seconds without explaining why he had barged his way in. He walked over to the unlocked filing cabinet next to Ganz's desk, searched through the well-organized records, and took possession of several documents. He told Madeleine Paqué and Georg Ising only that editor-in-chief Josef Ganz and editor Frank Arnau stood accused of blackmailing the German auto industry and that as soon as Ganz came home, he must report to the central police station. Madeleine cast him a malicious glance. She spent the rest of the afternoon nervously waiting for Josef, running to the window every time she heard a car coming. When he finally turned onto the drive in the early evening, she ran out and told him in a torrent of words what had happened earlier that day. As they talked, darkness came down over Frankfurt.

Charged with Blackmail

Early next morning, Josef Ganz reported to Frankfurt police headquarters, a large, rather grand building with a glass watchtower on the roof. There he was told by the officer on duty that he and Frank Arnau had indeed been charged by the Gestapo with blackmailing the German car industry. From the indictment, he concluded that the case had been put together by Ehrhardt and that it centered upon the clash between *Motor-Kritik* and Daimler-Benz three years earlier, in the spring of 1930, when the car manufacturer had covered the costs of a reprint in order to prevent the circulation of an issue containing unfavorable comments about the company. The page proofs of that unpublished issue had been seized as "evidence" by Ehrhardt during the search of the house. Ehrhardt claimed that Ganz had threatened to publish it unless Daimler-Benz paid a considerable sum,[1] and he went on to claim that Daimler-Benz had subsequently been forced to employ both Ganz and Arnau as consultants, paying them high salaries even though their jobs existed purely on paper. It was a grubby web of lies. The officer on duty told Ganz that he had until May 6 at the latest to respond to the charge in writing. Ganz returned to his office.

He knew it would be pointless to try to contact Frank Arnau, who had fled Germany deep in the night of April 1 using counterfeit documents.[2] The blackmail case fabricated by Ehrhardt was only one of a series of grave accusations made against Arnau, whom the Nazis had been hunting since the Reichstag fire, suspecting he was a member of the Roter Frontkämpferbund, the illegal paramilitary wing of the German Communist Party. With the support of the propaganda ministry and his wartime comrade Hermann Göring, minister without portfolio in the new government, Ehrhardt was trying to portray the activities of Arnau and Ganz as instances of Communist and Jewish conspiracy against the Third Reich. It was all part of a sensational series of stories collectively known as the Frank Arnau Affair,[3] but personal motives on the part of both Ehrhardt and Göring lay behind these latest events. To Ehrhardt, Josef Ganz was a damaging critic who must be eliminated, and Tatra had given him the task of laying claim to Ganz's Volkswagen design and patents. Hermann Göring was involved in a personal vendetta with Frank Arnau, who in 1932 had come forward as a crown witness against him in a case involving the embezzlement of RM 23,000 from the aircraft department at BMW.[4] When Arnau disappeared, BMW's director Franz Josef Popp, fearing the personal consequences, urgently requested that Arnau's secretary get rid of all paperwork relating to the charges BMW had brought against Göring, and when the Gestapo raided Arnau's house in Berlin in late April they found only a few smoldering scraps of paper in the oven and the heating boiler.[5] All the evidence had been destroyed, but as a result of pressure exerted as part of the Frank Arnau Affair, Popp decided to cut all ties with both of BMW's technical consultants, Frank Arnau and Josef Ganz. In the week after the house search by the Gestapo, Ganz received a letter from the directors of BMW terminating his contract after almost two years without further notice and with immediate effect:

> It is unfortunately necessary for us to cancel our contract with you as quickly as possible. We therefore terminate the agreement made with you on July 20, 1931, as of October 1, 1933. We wish to stress our appreciation of your services. As regards the completion of the work

Bayerische Motoren Werke
Aktiengesellschaft

Direktion

TELEGRAMM-ADRESSE: BAYERNMOTOR

FERNSPRECHER: Nr. 33737

Le.

MÜNCHEN 13, LERCHENAUERSTR. 76 3.Mai 1933.

Herrn

Diplom Ingenieur Josef G a n z ,

F r a n k f u r t a.Main,

Zeppelin Allee 85.

EINSCHREIBEN!

Betr.: Ihre vertraglichen Beziehungen zu unserer Firma.

Sehr geehrter Herr Ganz!

Wir sind leider gezwungen, unser Vertragsverhältnis zu Ihnen raschestens aufzulösen. Daher kündigen wir das am 20.Juli 1931 getroffene Abkommen zum 1.Oktober 1933.

Wir sprechen Ihnen hiermit ausdrücklich unsere Anerkennung für Ihre Dienste aus.

Bezüglich der Abwicklung der zur Zeit für uns in Angriff genommenen Arbeiten und Rückübertragung Ihres DRP Nr.528 643 bitten wir Sie, uns demnächst in München persönlich aufzusuchen.

Bei dieser Gelegenheit wollen wir auch der Frage nähertreten, Ihre Dienste späterhin von Fall zu Fall in Anspruch zu nehmen.

Hochachtungsvoll

BAYERISCHE MOTOREN WERKE
Aktiengesellschaft

Dismissal letter from BMW to Josef Ganz, 1933

> currently underway and the transfer back to you of your patent DRP 528 643, we request you to visit us personally in Munich soon. We would also like to take that opportunity to discuss whether we can use your services in future on a project-by-project basis.[6]

It was a huge disappointment to Ganz to be so suddenly dismissed by BMW, just at the point when the manufacturer was experiencing enormous success with the BMW AM1, developed under his consultancy, and with follow-up models also featuring independent suspension. With a mixture of regret and rage, Ganz gathered together all the evidence against Paul Ehrhardt he could find at his editorial office and wrote an extremely detailed statement that he delivered to police headquarters in Frankfurt, as required, on May 6. Later that afternoon, Paul Ehrhardt's car drew up. He asked the chief of police to give him Ganz's submission so that he could personally deliver it to Gestapo headquarters in Berlin. Ehrhardt put the brown envelope containing the statement and accompanying evidence in his briefcase and drove to Gustav Röhr's office at Adler. Röhr was a supporter of the accusations against Ganz. Ehrhardt updated him on the latest developments and asked whether perhaps he had a Wanderer Continental typewriter in his office, the make he had seen on Josef Ganz's desk and the machine used to type the statement. After searching for a while, Röhr discovered that Adler did have such a model. Ehrhardt took the first page of the statement out of the envelope, wound it into the typewriter, and typed the figure 2 in front of the date, so that May 6 now read May 26. Ehrhardt held on to the document for several weeks, so that it would seem to the Gestapo that Josef Ganz had failed to submit his statement on time.[7]

A Volkswagen for a Thousand Reichsmark

At his editorial office in Frankfurt, Josef Ganz calmly carried on getting the next issue of *Motor-Kritik* ready for publication, unaware of the dirty game played by Paul Ehrhardt behind the scenes. He did not want his work to be interrupted by enemy subterfuge at this exciting time in which the new National Socialist government had set itself the task of motorizing the German masses. The construction of a highway network announced by Chancellor Adolf Hitler and the Volkswagen that Ganz wanted to see built would be crucial to those efforts. The recently introduced Standard Superior represented a first step toward the Volkswagen, but Ganz believed its retail price of RM 1,590 was still too high to permit the sort of mass sales he had in mind. In early May, he wrote in *Motor-Kritik*:

> Average incomes in Germany in relation to the cost of living are still such that the popularization [of the Volkswagen] on a large scale is out of the question. People with reliable insight estimate that for a decisive increase in the number of cars the condition that needs to be met is more or less as follows: the purchase price, new, of a four-seater with a maximum speed of 60 km/h [37.2 mph], fuel consumption of 5 liters [47 mpg], a life expectancy of around 5 years, and access to cheap repairs must not exceed 1,000 Mark. [...] Is that possible? Certainly it's possible!

In Ganz's view, the key to the Volkswagen was to develop, using the most modern construction methods, a small car with four seats weighing only 350 kilograms (772 pounds). It would then be possible to keep the price down to RM 1,000, and the car would then fall within the class of vehicles that could be driven with a motorcycle license. At that price, Ganz envisaged a market in Germany for more than a million Volkswagens. He called upon "trend-setting circles" to get on with the task, rather than "dithering" for another ten years.[8]

This issue of *Motor-Kritik*, like many previous issues, must have landed on the desk of Chancellor Adolf Hitler, who had every interest in mass motorization and had promised at the latest IAMA in Berlin to mobilize Germany. He recognized the enormous potential propaganda impact of putting a Volkswagen on the market at a price of a thousand Reichsmark, and in the early summer he instructed his close personal friend and unofficial adviser in the field of auto engineering, Jakob Werlin, to research whether it was truly feasible to produce a car at that price. This was a practical follow-up to a conversation that had taken place in the summer of 1932, when, driving together through a heavy shower of rain, they had sprayed water and mud over a couple who passed them on a motorcycle. Hitler had said to Werlin: "By the devil it must surely be possible to build a car that costs no more than a powerful motorbike." He suggested that Werlin and the employees of Daimler-Benz should "rack their brains" on the matter. Werlin thought to himself: "That's easier said than done; we're not living in a totalitarian state!"[9] Now, a year later, the problem had landed in his lap again. Werlin contacted Popp, the director of BMW, who despite having studied the various options for several years was no different from any other auto manufacturer in seeing little to be gained by producing a Volkswagen costing a thousand Reichsmark. He told Werlin he thought it impossible.[10] Hitler, however, refused to relinquish the target retail price suggested by Ganz, a designer who wanted manufacturers to put their own interests aside and to see it as their duty as Germans to join forces and develop a Volkswagen, whether the government supported the project or not.

Ganz's dream was that a collaborative initiative by industry and government would lead to his being invited to join a consortium tasked with developing the German Volkswagen. He followed up his call for a maximum retail price of RM 1,000 by working night after night in the early weeks of May on a long article setting out his Volkswagen design in detail. He included his experiences with the Ardie-Ganz and Adler Maikäfer prototypes, the story of the Standard Superior, and detailed descriptions of his design, illustrated with drawings from his patents. He wrote that a vehicle with a rear-mounted engine and streamlining was "undoubtedly the optimal solution to the passenger vehicle problem." Such a rear-engined car might also remove resistance to streamlining, since "the organic fusion" of a chassis with an engine at the back and streamlined bodywork would lead to an uncompromising, harmonious whole of "noble functional beauty."

Ganz regarded the switch by several auto manufacturers to the production of front-wheel-drive models over the past few years as merely a sideways step, away from the old-fashioned way of building cars with long drive shafts but no more than a brief dalliance on the route to the streamlined, rear-engined car. He was convinced the National Socialist government would bring an end to this "developmental lag." Ganz had the kind of clear-cut vision and passion manifested by designers at the forefront of modernist movements like the Bauhaus in Germany and De Stijl in the Netherlands; instead of following fickle fashion, he had based his Volkswagen design on a purely scientific approach to streamlining, the epitome of "form follows function." That month, Ganz applied for a patent on a pure, streamlined "beetle shape," with the lines of the roof and rear fenders meeting at a single point.[11] For full access to the depths of the engine compartment, the entire back shell of the car hinged up. The article ran to a full seventeen pages of text and illustrations. An affectionate Josef Ganz used as his opening photograph one of the pictures taken two years earlier of little Dieter, Madeleine Paqué's nephew, playing on the floor with a model of the Maikäfer along with

Klasse 63 c_2. **Ausgegeben am 10. November 1936.**

ÖSTERREICHISCHES PATENTAMT.

PATENTSCHRIFT Nr. 147521.

DIPL. ING. JOSEPH GANZ IN FRANKFURT A. M.

Stromlinienförmiger, die Räder auch an den Außenseiten umschließender Wagenkasten für vier- und mehrrädrige Kraftfahrzeuge.

Angemeldet am 12. Mai 1933. — Beginn der Patentdauer: 15. Juni 1936.

Header of AT patent #147521, for a streamlined, beetle-like car, first filed by Josef Ganz in Germany, 1933

Dolly the German shepherd. Entitled "For the Small Man's Car," his magnum opus was published in *Motor-Kritik* number 10 of 1933.[12] In his opening address, Ganz spoke enthusiastically of Hitler as the "Führer of the German people" who on May 1 had announced his government's plans to "repair" the German economy, which included an ambitious program of road building. Josef Ganz was greatly in favor of this "support of the economy by the state" and called upon automobile engineers "to build cars that match the living standards of German citizens." He stressed that it was "more important than ever" to stimulate "the development of the German Volkswagen." Hitler's determination to support his lifelong dreams had perhaps blinded Ganz to the darker sides of the Nazi regime. He still experienced his recent problems as personal slander by people like Paul Ehrhardt and not as the result of active anti-Semitic government policy. Like so many assimilated German Jews, he could not foresee the devastating consequences his Jewish origins were to have. Thousands of copies would be distributed in Germany and far beyond in mid-May. The blueprint for the Volkswagen was complete.

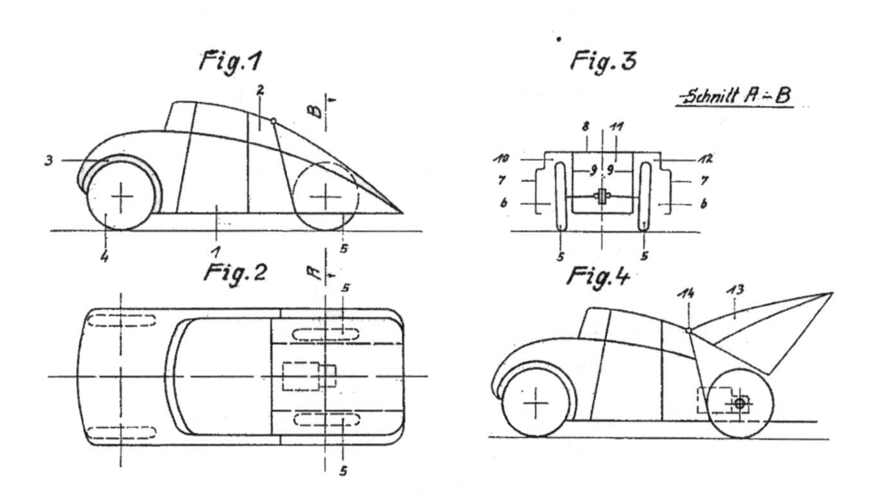

Drawing from AT patent #147521, for a new, streamlined, beetle-like car, first filed by Josef Ganz in Germany, 1933

Grand Prix on the AVUS

Josef Ganz was so busy he hardly had time to think about the absurd charges brought against him of blackmailing the German car industry. Two weeks had passed since he had submitted his written statement to police headquarters, and he had heard nothing since. Now he was about to leave for Berlin to watch the Grand Prix due to be held on May 21 on the Automobil-Verkehrs- und Übungs-Straße (AVUS), a stretch of road that had been adapted into a racing circuit. He wanted to be there as early as possible to watch the racing models from far and wide being delivered and carefully unloaded. A truck had been brought in purely to transport the dozens of tires that would be worn out during the race. He systematically photographed the whole process, as well as all the special features of the vehicles, enjoying close-up views of racing cars from prestigious companies like Bugatti, Delage, and Amilcar in France, Alfa Romeo in Italy, and MG and Austin in Britain. He spoke reasonably good English and French and enjoyed talking with foreign participants about their cars and their racing performances.

The Grand Prix on the AVUS was a major event, and like all other prominent happenings in Germany in those years it was swallowed up by the Nazi propaganda machine. Huge swastika flags flew, SA troops marched along nearby streets, and many Nazi Party bigwigs attended. Among those on the rostrum were the ministers Joseph Goebbels, Hermann Göring, and Alfred Hugenberg. The representatives of the Third Reich found it deeply troubling that the faster laps were being driven by foreign racing cars from Bugatti, Alfa Romeo, and Delage, while German companies Mercedes-Benz, BMW, and DKW trailed behind. To ensure Germany achieved absolute victory on the racing circuits of Europe, the government had recently provided financial support for the development of new racing models by Auto Union and Daimler-Benz. The car that Auto Union intended to build was designed by Ferdinand Porsche, with a mid-mounted engine and swing axles like the Tropfenwagen, but it was still very much at the design stage. Daimler-Benz, by contrast, had wanted to race its new streamlined Mercedes-Benz SSK on the AVUS, but three days before the official start, its driver Otto Merz rolled the car during a test run on the circuit and was killed. His death overshadowed the Grand Prix, but it did little to detract from the participants' competitive spirit, and there was immense public interest. Spectators arrived in Berlin in droves.

As editor-in-chief of *Motor-Kritik*, Josef Ganz had a seat reserved for him on the closed press stand, and he could walk freely among the cars in the pit stop. This was where he felt at home, amid the roar of powerful engines with unmuffled exhausts, the gasoline fumes, the stench of oil, and the chaos of busy mechanics. It was here that he could meet the most important experts, engineers, and drivers. Participants and visitors smiled at him as he took pictures of the crowded stands.

Arrest in Berlin

Everywhere on the AVUS were newspaper and magazine journalists, cameramen making cinema newsreels, and radio reporters. Josef Ganz would never get the chance to report on this Grand Prix for *Motor-Kritik*. Later in the day, when he tried to return to his hotel room, he found two sinister-looking men in long black leather coats waiting for him in the hotel lobby. They turned out to be Gestapo agents who had been tipped off by colleagues in Frankfurt that they would find him there. After a brief exchange of words, he was arrested and driven away in a black car, which raced through the center of Berlin in the twilight to police headquarters on the Alexanderplatz. Since its founding less than a month before, the Gestapo had occupied this imposing red-brick building with copper-sheathed towers, popularly known as the Red Fortress. From the main entrance, the two Gestapo men marched Ganz along endless corridors, deep into one of the wings of the building. There he was registered as a prisoner, his personal belongings were taken away, and he was led to the cells in the basement. The large communal cell into which he was flung was already overcrowded, and he sat for hours surrounded by dozens of Communists and other "enemies of the state" in the chilly, damp cellar, without even sufficient room to lie down and sleep.[13]

After a long night spent dozing in a seated position on the ground, Ganz was fetched by a guard the following morning and taken to an interrogation room. During his long grilling, he was forced to listen time and again to the same set of fabricated accusations, which originated with Paul Ehrhardt. He could do little other than repeatedly refer to the written statement he had delivered to the police in Frankfurt two weeks before. According to the Gestapo, it had never arrived in Berlin. Because Gestapo headquarters in Frankfurt and Berlin were not yet integrated, Ehrhardt was able to hold on to the evidence for weeks. The interrogators in Berlin therefore had no reason to believe Ganz was telling the truth. The interrogation lasted several days. Every morning, after a long, exhausting night, he was taken from the communal cell, questioned in a similar manner, and after many hours led back again.

News of Josef Ganz's arrest spread around the industry like wildfire, to judge by the many expressions of support that flooded into the editorial office from friends, acquaintances, and readers of *Motor-Kritik*. The telephone never stopped ringing, and the mail piled up. In those first anxious days, Madeleine Paqué and Georg Ising found themselves unable to contact Josef, and they were refused admittance to the Red Fortress. After three days, they finally received good news in a phone call from Baron Klaus-Detlev von Oertzen, a close acquaintance of Josef who worked for the auto manufacturer Wanderer and who had supported him during the smear campaign in the Nazi magazine *Die Nationale Front* by testifying against Paul Ehrhardt. Hearing the incredible news of Ganz's arrest, never doubting the accusations were groundless, he had rung his old wartime comrade Hermann Göring and convinced him that the blackmail case must be based on false evidence. He had known Josef Ganz for years, he said, and it was out of the question that a single word of the charge could be true. Von Oertzen strongly advised Göring that the Nazi Party should withdraw the allegations to avoid discrediting itself. Göring took his advice seriously and decided to transfer the case from the Gestapo to the public prosecutor.[14] By then, Ganz had spent three days in a Gestapo cell, with almost no sleep and unable to wash or shave. Under the circumstances, it was a great relief to him to hear from the guard that the charge had been handed over to the public prosecutor and that he would be moved to an ordinary one-man police cell. He was even allowed a brief phone call. He immediately rang his beloved Madeleine in Frankfurt, who was very emotional but happy to hear his voice. Ganz reassured her and said he was certain he would be released very soon. He instructed his colleague Georg Ising to take over as acting editor-in-chief until his return and to prepare the next issue of *Motor-Kritik* for the printer. In that issue, Ising included a short item about Ganz's arrest:

> Frank Arnau (Heinrich Schmitt) is being investigated on extremely serious charges (corruption, extortion, shady currency deals, tax evasion etc.). Frank Arnau is abroad. In connection with the investigation into Frank Arnau, others including our *Diplom-Ingenieur* Josef Ganz are in custody. The [editors of] *Motor-Kritik* and all those who know Josef Ganz are convinced of his irreproachable behavior and hope he will quickly be freed and exonerated. Herr Ganz himself, after the subterranean battle of recent years, welcomes the forthcoming judicial clarification and requests all his *Motor-Kritik* friends to be patient in the meantime.[15]

That "subterranean battle," which had been joined over recent years by the Reichsverband der Automobilindustrie (RDA), among others, was far from over. In its determination to destroy *Motor-Kritik*, the RDA took the opportunity presented by the Frank Arnau Affair to accuse Ganz once more of "damaging the German economy," claiming that articles in *Motor-Kritik* had caused models to date so quickly that German manufacturers were left with thousands of superseded and unsalable cars, worth RM 350 million in total.[16] This new charge led to an extension of Ganz's preindictment custody. He was taken out of his police cell in the Red Fortress and transferred to prison on the Lehrter Straße in the Berlin district of Moabit, a name synonymous with jail. The building was directly behind the imposing complex of the Moabit criminal court, the largest court building in Europe, decorated with vast pillars and pilasters, huge statues, and eagles carved out of sandstone. The Moabit prison was a star-shaped building with five wings and a central dome on the Benthamite model. Long rows of single cells on several levels lined the wings on either side of an open, cast-iron stairwell. In Moabit, Josef Ganz was in "good company," as he called it: the ex-mayor of Berlin occupied the cell next to his.[17]

Libel against Ganz and Arnau

In other media too, articles were now appearing about the corruption case against the fugitive Frank Arnau. The *Frankfurter Zeitung* called Arnau a "versatile swindler-genius" and "one of the most unscrupulous of crooks," in whose house in Berlin a search had revealed "unimaginable things." Arnau was said to have made thousands of Reichsmark by means of "large-scale blackmail maneuvers." He had been earning, according to the paper, no less than RM 150,000 a year. The article went on:

> He was aided in this by an engineer and writer who has been arrested in Berlin. Together these two gentlemen published extra issues of a motoring magazine, in which they concentrated entirely and exclusively on so-called revelations about wrongdoing in certain factories in the auto industry. Many firms, who were of course informed about these extra issues in good time before they appeared, often paid sums of money purely to prevent the good name of their products being associated with this kind of grubby discussion—even if the accusations were libelous.[18]

Shortly after his arrival in Moabit, Ganz received an unexpected visit from the Berlin lawyer Dr. Thom, who told him that his sister Margit and her ex-husband Jakob Feldhammer had read of his arrest in the papers. Margit had contacted Prince Erbach, the German ambassador in Vienna, an acquaintance of hers from the time their father Hugo Ganz worked as an editor at the *Frankfurter Zeitung*. Feldhammer had rung his sister

Technical director Hans Ledwinka (left) and Baron von Ringhoffer of the Ringhoffer-Tatra company at the Karl Benz memorial event, Mannheim, 1933. Shortly after this event, Josef Ganz would be arrested on falsified charges brought by Paul Ehrhardt, employee of Tatra and Gestapo agent, in an attempt to lay claim to Ganz's Volkswagen design for the Ringhoffer-Tatra company.

After a few weeks of solitary confinement in Berlin's Moabit prison, June 1933, Josef Ganz was released. He fetched his belongings, including his camera, and took a few candid shots of the institution and its personnel. This photo shows the long rows of single cells.

Josef Ganz secretly took a photo of the guards in Moabit prison as they handed him back his belongings, Berlin, 1933. The standing guard got wise just as Ganz clicked the shutter.

Dr. Thom, Josef Ganz's lawyer, standing in the street outside of Moabit prison, shortly after the release of Ganz. Next to him are two suitcases and a briefcase full of evidence, Berlin, 1933.

A visibly tired Josef Ganz shortly after his release from Moabit prison, Frankfurt, 1933.

Next page: Shortly before the arrest of Ganz, production was started on the Standard Superior. This innovative car attracted a lot of attention on the streets of Germany, 1933.

Printz

Top: Official dealer introduction of the Standard Superior, 1933.
Bottom: The Standard Superior competed in various car rallies, including a 2,000 km race crossing Germany, 1933. By the time Josef Ganz was released from prison, such events were flooded with Nazis, showing off their brown uniforms.

Top: A couple of Standard Superiors are being prepared at the factory to compete in the 2,000 km rally crossing Germany, Ludwigsburg, 1933.
Bottom: A Mercedes-Benz competing in the 2,000 km car rally crossing Germany, 1933.

Closing ceremony of the 2,000 km rally, honoring winner Adler Trumpf, Baden-Baden, 1933. These photos by Josef Ganz also document the unavoidable Nazification of Germany.

HEINRICH SCHREIBER
219

Above: Josef Ganz, wearing a nametag with little swastikas, at the small-vehicle competition, Heppenheim, 1933.
Left: Standard Superior competing in the small-vehicle competition, 1933.

Next page top: The Standard Superior (left) in between other competitors of the small-vehicle competition, mostly three-wheeled cars and motorcycles, 1933.

Above: Two three-wheeled Framo Stromers in the small-vehicle competition, 1933.

Middle: Local children point the right way to Josef Ganz driving up in his Tatra, 1933.

Factory driver Otto Wegmer steered the Standard Superior at the small-vehicle competition, Heppenheim, 1933. A competitor challenged Wegmer to corner his car on a gravel path, in tighter and tighter arcs, until the thin tires could no longer cope. Centrifugal force flipped the Standard Superior over, but Wegmer and the car remained unharmed.

Advertising inside the spare wheel on the back of the Standard Superior. Due to extra investment for a new body design for the production model, the company was forced to raise the price from 1,590 to 1,620 Reichsmark, 1933.

Anna, who was married to the German steel industrialist Victor Hoesch, and Hoesch had engaged Thom to help Ganz.[19] Thom had immediately begun studying the case, but to his regret he had to tell Ganz that no progress had been made because the documents held by Paul Ehrhardt had still not arrived in Berlin. Two weeks went by while the case was under investigation, and the public prosecutor tried to assemble all the evidence, including Ganz's statement, held back by Ehrhardt. Acting editor-in-chief Georg Ising and *Motor-Kritik*'s publishers Hartwig and Wilhelm Breidenstein meanwhile did all they could to help Ganz by gathering evidence against Ehrhardt. Hartwig Breidenstein even contacted the police in Davos, Switzerland, to obtain details about charges dating back to 1919 that involved cheating at poker and deceiving a factory director.[20] Ganz was able to provide plenty of witness statements hostile to Paul Ehrhardt and Gustav Röhr, garnered over the past two years. The evidence against Ehrhardt and his fabricated blackmail case was piling up. The editors also received letter after letter from readers, expressing support for the work of Josef Ganz and *Motor-Kritik*. By the time the next issue was due to go to the printer, Georg Ising had been functioning as editor-in-chief for almost three weeks. *Motor-Kritik* number 12, the mid-June issue, opened with the announcement:

> *Motor-Kritik* is on everyone's lips, because Dipl.-Ing. Ganz has been implicated in the Frank Arnau Affair. Readers of *Motor-Kritik* know that this magazine expresses harsh criticism, perhaps sometimes excessively harsh, but certainly none that serves no purpose.[21]

In an accompanying article, Ising underlined the importance of criticism and freedom of expression:

> *Motor-Kritik* is intended above all for the specialist (as is adequately demonstrated by its choice of subjects and their treatment), and it presupposes the intellect of the independently minded and critical professional who is perfectly able to tell where the boundary between seriousness and satire lies. Most important of all are the repercussions of its

> criticism, the aim being to assess a subject in such a way as to identify opportunities for improvement. [...] Critical opinions in *Motor-Kritik* have had a progressive effect; this cannot be disputed, since countless suggestions made here have been implemented over time, and implemented by those who at first adopted an attitude that was not only negative but, in a word, hostile. That, after all, was the ultimate goal.[22]

"Justified criticism," Georg Ising wrote, does not allow itself to be suppressed, "even when it contradicts a hundred times over the rose-tinted statements emanating from an advertising department. A critic can be silenced, criticism can't." Because of Ehrhardt's attempts to silence him, the critic Josef Ganz had now been locked up for over three weeks. The case began to make headway again only when the documents held back by Ehrhardt were at last produced under considerable pressure from the public prosecutor in Berlin. After a stay of a month in the Moabit jail, Josef Ganz appeared before the prosecuting attorney on June 26. A guard escorted him down a long subterranean corridor from the cells to the court building and on via another network of corridors to the courtroom. The judge was a surly older man who even now tried to prove Ganz guilty. The hearing was tense and did not last long; the judge took one of the pieces of evidence provided by Ganz and said: "Do you admit that this solemn statement by Dr. Nibel is a fake?" Irritated by the way things were going, Ganz answered: "How can you allege such a thing? Simply phone Dr. Nibel if you doubt its authenticity!" The judge ordered Ganz to be removed from the courtroom, and he was taken back to prison. Without any further word, he spent yet another night in his tiny, bare cell. The guard arrived the next morning and unlocked the heavy steel door, which swung open to reveal a smiling Dr. Thom, Ganz's lawyer, who declared triumphantly: "You're free!"[23] Josef Ganz left his cell along with Thom to fetch his personal belongings, including his camera, which still had an undeveloped roll of film in it with pictures of the Grand Prix on the AVUS taken more than a month before. He took a few shots inside and outside the prison of the guards, the cells, and Dr. Thom standing in the street outside with two suitcases and a briefcase full of evidence beside him.

Back in Frankfurt

Dr. Thom drove Ganz to Berlin's Anhalter railway station, where he rang Madeleine Paqué to tell her the good news of his release and to say he was taking the first available train back to Frankfurt. Many hours later, Madeleine stood on the platform of Frankfurt's main station, watching excitedly as the black-and-red steam train from Berlin approached and stopped with a loud hiss. The doors to the compartments swung open, and Madeleine searched for Josef in the human throng that poured out. She soon spotted him in his rather crumpled suit, holding the little travel bag with which he had left a month earlier. She was delighted to see him, and together they walked out to where Georg Ising was waiting in the Tatra to drive them home.

The following day, Ganz and Ising went to visit Hartwig Breidenstein, the publisher of *Motor-Kritik*, who was extremely glad to see Ganz. He looked concerned, however, and said that the previous day he had gone to Jakob Sprenger's office in Darmstadt, at Sprenger's insistence. Sprenger was Reichsstatthalter (the term for a provincial ruler under the Nazi regime) of the German state of Hesse, and he told Breidenstein that *Motor-Kritik* was nothing more than a sensation-seeking rag with a Jewish editor-in-chief against whom serious charges had been brought. The blackmail case was still ongoing, and Sprenger said that Gustav Röhr of Adler had submitted a fresh accusation against Ganz, claiming there had been an attempt at extortion during negotiations over royalties for the building of the Maikäfer. He neatly passed over the fact that Röhr had not even been working for Adler at the time. Sprenger had put Breidenstein under heavy pressure to sack the Jew Ganz as editor-in-chief and to publish only a brief message in the magazine about his release, without any further comment.[24] Breidenstein said with great regret that he had little choice but to obey. He did promise to let Ganz continue carrying out important work on a freelance basis. The very next day, a letter arrived from the publisher, confirming that the contract had been terminated "by mutual agreement," as of three months hence.[25] After five years, Ganz's name was silently removed from the pages of the magazine.

H·BECHHOLD VERLAGSBUCHHANDLUNG
INHABER: F. W. UND W. BREIDENSTEIN · FRANKFURT-MAIN · BLÜCHERSTR. 20-22

H. BECHHOLD VERLAGSBUCHHANDLUNG, FRANKFURT-M., BLÜCHERSTR. 20-22

BANKKONTEN: FRANKFURTER GENOSSENSCHAFTSBANK, DRESDNER BANK FRANKFURT-M., AMSTERDAMSCHE BANK AMSTERDAM, DEUTSCHE AGRAR- U. INDUSTRIEBANK PRAG ● POSTSCHECKKONTEN NUMMER 35 FRANKFURT-M., POSTSPARKASSENAMT WIEN 79258, PRAG 79906, ZÜRICH VIII 5926 ● FERNRUF: SAMMEL-NR. SENCKENBERG 30101 ● DRAHTANSCHRIFT: UMSCHAU FRANKFURTMAIN

ZEITSCHRIFTEN:
DIE UMSCHAU
MOTOR-KRITIK
HALBMONATSSCHRIFT FÜR DEN VERBRAUCHER VON KRAFTFAHRZEUGEN
BÜCHER DER UMSCHAU
HANDLEXIKON DER NATURWISSENSCHAFTEN UND MEDIZIN u. a.

Herrn
Dipl.-Ing. Josef Ganz

Frankfurt a/M.
Zeppelin-Allee 85

Ihr Zeichen	Ihre Nachricht vom	Unser Zeichen	Haustelefon-Nr.	FRANKFURT A. M., BLÜCHERSTR. 20-22
		BV.Br/We.	12	29. Juni 1933.

Sehr geehrter Herr Ganz!

Wir bestätigen hierdurch unsere mündliche Vereinbarung dahingehend, dass der im § 13 unseres Vertrags vom 27. Februar 1929 festgesetzte Kündigungstermin vom 30. Juni 1933 ausnahmsweise infolge der allgemeinen Verhältnisse auf den 30. September 1933 in beiderseitigem Einverständnis verlegt wird, unter Kürzung der Kündigungsdauer auf drei Monate. Es ist selbstverständlich, dass wir die Zwischenzeit zur Klärung unserer weiteren gemeinschaftlichen Arbeit benutzen werden und bitten Sie der Ordnung halber um eine kurze Gegenbestätigung.

Mit freundlichem Gruss

Dismissal letter from *Motor-Kritik*'s publisher to Josef Ganz, 1933

Georg Ising stressed in the subsequent issue how great the contribution of *Motor-Kritik* had been to the recovery of the German auto industry and how much opposition the magazine had encountered over the past few years. He deliberately avoided naming names. Below his article was a short, neutral item by Josef Ganz, who wrote:

> In this position I naturally feel compelled to account for everything. Unfortunately I must refrain from doing so, to avoid obstructing the progress of any further investigation. All I venture to say in so many words now is that no one who has placed faith in me and in *Motor-Kritik* will be disappointed. My profound thanks to all the friends and colleagues who have mobilized on my behalf and unselfishly offered me their help.[26]

Shortly after the publication of this latest issue, Ganz received a telephone call from Professor Doctor Carl Heinrich August Lüer, president of the Frankfurt Chamber of Commerce, whom Hitler had appointed Treuhänder der Arbeit for the state of Hesse, a kind of mediator in conflicts between workers and management. He had heard that the Gestapo had arrested Paul Ehrhardt for falsifying evidence but had been forced to release him immediately, since Ehrhardt enjoyed protection as a secret agent for the SD. Lüer had also been in direct contact with Reichsstatthalter Jakob Sprenger and had heard that Paul Ehrhardt, Gustav Röhr, and the car manufacturer Bernhard Stoewer were making efforts behind the scenes to persuade the parties to have Ganz arrested again and taken to the recently opened Dachau concentration camp. Lüer secretly kept in close touch about these developments, while the investigation continued.[27] Ganz waited for the right moment to strike back.

CHAPTER ELEVEN

VOLKSWAGENS FOR THE AUTOBAHN

That summer, little appeared to have changed at the editorial offices of *Motor-Kritik*. Josef Ganz and Georg Ising worked together closely to maintain the progressive character of the magazine. Officially, however, Ising was now editor-in-chief, and Ganz, the founder of *Motor-Kritik*, had been demoted to the position of editor, in response to pressure from the Nazi Party. To escape the oppressive heat of the office, Josef Ganz, Georg Ising, and Madeleine Paqué often withdrew to the large garden behind the villa, where the German shepherd Dolly, with her thick coat of fur, lay panting in the shade. In those idyllic, peaceful moments in the beautiful, burgeoning garden, Ganz spoke with enthusiasm about the decisive measures the authorities had put in place to motorize Germany. He had now been working for ten years to develop a Volkswagen, and, finally, there was a government in Berlin that understood its importance. Small cars like the Standard Superior, based on his patents, could now be driven by anyone with a motorcycle license, and the building of a new network of intercity highways had recently been announced.

Also that summer, a start would be made on the first stretch of Autobahn, covering the sixty-odd miles from Wiesbaden to Heidelberg via Frankfurt and Mannheim. Ganz was convinced that the new highway network would accelerate the introduction of streamlined vehicles large and small, with rear-mounted engines and independent suspension, since an open road would increase the demand for fast and efficient cars. Proof of the strength of his design concept was provided on July 8, 1933, by Freiherr Reinhard von König-Fachsenfeld, a racing driver and the German agent for Paul Jaray's streamline patents. He set five new world records that day on the French Autodrom racing circuit in Montlhéry under the official supervision of the Automobile Club de France, in a streamlined sports car based on the Standard Superior with a 0.5 liter engine, achieving an average speed of almost eighty miles per hour.[1] In *Motor-Kritik*, Von König-Fachsenfeld wrote that good road holding no longer depended on the weight of the vehicle and described the minimal engine capacity needed to make the Standard-Sport, freed of "all burdens, whether of weight or resistance," run at 120 to 140 kilometers per hour (seventy-five to eighty-seven mph).[2] Ganz responded in the magazine that he hoped the "dream of a people's vehicle" based on the Maikäfer could now become a reality and that "finally the combined forces that have worked against it for years and up to now have regarded no means as beneath them" had finally met their match.[3] The blueprint for such a "people's vehicle" had been described in detail in *Motor-Kritik* two months earlier, and the Standard Superior was daily proving the power of the concept in practice. Months after its introduction at the Berlin motor show, the press was still publishing articles praising both Ganz and the car based on his patents and ideas. The renowned specialist magazine *Allgemeine Automobilzeitung*, for example, declared:

> That which until recently was the focal point of the clash of opinions has now become a reality: the driving machine with two seats and the power of a large car. Many obstacles and prejudices had to be overcome before the Standard Superior could arrive on the scene, turning tumultuous and tenacious propaganda into material and form. For years its designer, Dipl.-Ing. Ganz, has promoted swing axles with parallel steering rods, independent half-axles for the drive wheels, a rear-mounted engine, and streamlining.[4]

Retail Price Too High

The magazine *Motor und Sport*, which had published a test-drive report on one of the first preproduction models of the Standard Superior in April 1933, came out with a second test summary in July, this time written by its editor-in-chief Wolfgang von Lengerke. He said that the fact that the first test with the not yet fully finalized production model had been so favorable should be regarded "as remarkable evidence of the capacities of the design, which has now been further perfected in essential details." The Superior was the smallest German car, Von Lengerke wrote, "but as far as its design goes the most advanced despite its small proportions."[5] *ADAC-Motorwelt*, the official organ of the German car club ADAC, praised Ganz's pioneering work:

> This structurally very interesting model has emerged from a prototype design by Dipl.-Ing. Josef Ganz of Frankfurt, who made his name as an advocate of small cars. The Standard Superior, as devised by Ganz, has proven extraordinarily successful—a fact attributable to the years of testing of the said research design—and it will in all probability lead to the introduction of larger models with rear-mounted engines. A second rear-engined, all-purpose car, designed by Dr.-Ing. Porsche, is known to be at the preproduction stage.

ADAC-Motorwelt was referring to the Zündapp Type 12 prototype, for which Porsche had recently sought publicity even though Zündapp had terminated the project many months before. "The only real shortcoming of the Standard Superior," the ADAC's magazine went on, was the relatively high retail price of RM 1,620, which stood in the way of the popular success of a car "such as Ganz has in mind." This did not mean, it explained, that the Standard Superior was too expensive in an absolute sense, but the price was so close to that of the two cheapest models from Opel and DKW "that the difference of around 300 to 350 Reichsmark will in many cases be insufficient to win those interested over to the Standard Superior."[6] Two months earlier, Josef Ganz had drawn attention to this problem in *Motor-Kritik*. He advocated a retail price of no more than RM 1,000 for a four-seater Volkswagen with a maximum speed of sixty kilometers per hour (thirty-seven mph) and consumption of five liters of gasoline per hundred kilometers (forty-seven mpg).

"The precondition for the successful motorization of Germany is connected to the solving of a technical problem, namely: lightweight cars," wrote Josef Ganz in *Motor-Kritik* in early August. "The perfect light car has a key part to play in the motorization of Germany. But how do we create the German Volkswagen now? How do we create it as quickly as possible?" The solution, he said, did not lie in engaging one particular car manufacturer but in bringing together the best professionals from all across the German auto industry. He appealed to them: "No German structural engineer should stay in the background if he is called upon to serve this end."[7] Ganz was himself eagerly waiting to be commissioned to develop the optimal Volkswagen as part of a government-financed consortium of top engineers. In that same issue of *Motor-Kritik*, he published a drawing of a design for a streamlined four-seater Volkswagen with a rear-mounted engine and swing axles.[8] The call never came. As Jakob Werlin had discovered when researching, on Hitler's behalf, the opportunities for designing and building a Volkswagen, the established auto manufacturers were not the least bit interested in openly sharing their trade secrets with others in a design consortium, least of all for that purpose.

2,000 Kilometers through Germany

Meanwhile, the Standard Superior was becoming increasingly famous. In July 1933, the factory racing drivers H. E. Schumacher and Otto Wegmer of the Standard Fahrzeugfabrik drove two Standard Superiors with 0.5 liter engines in the car rally 2000 KM Durch Deutschland (2,000 kilometers through Germany), a 1,243-mile race organized in part by the NSKK. Lined up at the start of this large-scale comparative test, alongside the little Standard Superior, were motorcycles and cars of various classes from firms including Adler, DKW, Horch, Mercedes-Benz, and Wanderer. Everywhere NSKK men could be seen striding about in uniform, with swastikas on their sleeves. Otto Wegmer drove the little Standard Superior for 1,297 miles nonstop, averaging thirty-seven miles per hour, without requiring repairs or any other assistance. The car had the smallest engine capacity, yet it won an impressive victory in its class, which included all vehicles with up to one-liter engines.[9] The overall winner was Adler, with its front-wheel-drive Adler Trumpf, developed by the engineer Joseph Dauben, to which elaborate tribute was paid at the closing ceremony in Baden-Baden. The city center had been decorated for the occasion with blood-red flags and banners bearing jet-black swastikas. Josef Ganz walked freely among them. As a journalist for *Motor-Kritik*, he followed and photographed the rally from start to finish. With some surprise, he watched the crowds giving the Hitler salute while the band was playing. A recent government decree had made the salute mandatory for all German citizens during the singing of the national anthem and the *Horst-Wessel-Lied*, the anthem of the Nazi Party. As everyone stood in the central square, right hands in the air, Ganz's right hand was on his camera, clicking the shutter to record the historic moment.

When the rally was over, the winning Standard Superior drove straight on to the next, the 1000-km-Fahrt, or 621-mile race, without undergoing a single repair. Ganz asked Standard director Wilhelm Gutbrod to loan him the second Standard Superior, which had only just been returned to Stuttgart after the 2000 KM Durch Deutschland. In early August, with the number 169 still emblazoned on the door, he drove to the Feldberg, a mountain in the middle of the Black Forest in the southwest of Germany, for a meeting of progressive auto engineers, all of them friends of *Motor-Kritik*. Among those he had arranged to meet was Paul Jaray, who had driven from Switzerland in a streamlined prototype he had developed five years earlier for the American car manufacturer Chrysler. Just before he reached the Feldberg, the front left side of the car hit a cyclist who had come down a hill at great speed. The man rolled "like a hedgehog" over the hood, along the car, and onto the ground. To Ganz, this was absolute proof that streamlining was not only more efficient but safer. "The smooth streamlining was his good fortune," he wrote later in *Motor-Kritik*. "From the way the incident turned out, we can conclude that the seriousness of traffic accidents, which are simply unavoidable, will be reduced by the introduction of streamlined cars."[10] While an old man in the village applied filler and paint to Paul Jaray's car, Ganz took photos and started talking to him. He turned out to have worked at the Zeppelin factory where Jaray had developed his ideas about streamlining years earlier and tested them in the wind tunnel. In beautiful sunny summer weather, smiling children watched, and the little village on top of the Feldberg was an oasis of calm in a turbulent time, a remote spot not yet penetrated by the swastikas of rapidly advancing Fascism.

Ganz and Jaray sat on a small terrace and talked excitedly about opportunities for building a streamlined automobile based on Jaray's patents, with a chassis modeled on that of the Standard Superior, a car matching the description Ganz had published in *Motor-Kritik* two weeks before. It would be the basis of the ultimate German Volkswagen. It was late evening by the time Ganz left the Black Forest. When he returned the Standard Superior to the factory in Stuttgart the next day, Wilhelm Gutbrod gave him the good news that as a result of a new initiative the Superior would have a chance to measure itself against all the other small cars manufactured in Germany. The ADAC, in collaboration with the NSKK, was planning to organize a small-vehicle competition, or *Kleinstkraftfahrzeugbewerb*, on August 15-17, pitting minicars and motorcycles of German manufacture against one another. The aim was to compare all small vehicles, with a view to developing a German Volkswagen. Both Ganz and Gutbrod were more than eager to take part, Gutbrod saying he was convinced the Standard Superior would achieve a glorious victory. This would be the perfect chance to prove, in a public and neutral arena, the enormous advantages of the Maikäfer approach to car building.[11]

The Quest for the Volkswagen

In the early morning of August 14, 1933, Josef Ganz set off in his Tatra for Heppenheim in central Germany, where the small-vehicle competition was to be held. Heppenheim, a historic town with a Jewish community that could be traced back to the Middle Ages, lay amid beautiful mountainous landscapes with good provincial roads. In these idyllic surroundings stood the large Heppenheim psychiatric institution. Local residents were not at all happy that "lunatics" from all over that part of Germany were locked up in their town. Josef Ganz joked about it with his colleague Walter Ostwald, who lived in Heppenheim and was one of the organizers of the competition—"Walter Ostwald is among the Heppenheimers who are allowed to go about freely. That surely bodes well for him"—but the Nazis regarded lunatic asylums as a millstone around their necks, an unnecessary expense, and a few years later the psychiatric institution in Heppenheim would be emptied under the Nazi euthanasia program. In that beautiful summer of 1933, the little German town was exceptionally peaceful. The only signs of the political changes brewing even there in the scorching heat were the small, discreet swastikas on the pennants affixed to the cars and on the name tags pinned to the jackets and shirts of participants. Josef Ganz walked around with a name badge bordered by swastikas.

Cars and motorcycles came and went, dozens at a time. Every conceivable small vehicle was here: motorbikes, motorized carrier tricycles, three-wheeled cars, and of course the Standard Superior—the only small four-wheeled car with a rear-mounted engine—which Ganz photographed many times. The little "beetle" zoomed flawlessly through all its tests and trials. At the wheel again, this time bearing the number 61, was factory driver Otto Wegmer, fresh from his victory in the 2000 KM Durch Deutschland. Its most direct competitor was the DKW Front, a four-wheeled, front-wheel-drive car, of which five were competing. The rest were all three-wheelers: one Mayrette, two Hercules, three Goliath Pioniers, and four Framo Stromers. The competing vehicles were divided into eight groups, depending on their type and the cylinder capacity of their engines. There were three groups for motorcycles, three for small cars, and two for small commercial vehicles. In between races, all the cars and motorcycles stood on a large parking lot in the center of Heppenheim, in the shade of tall trees.

As the Feldberg had been a week or two earlier, Heppenheim was a peaceful oasis to Josef Ganz. The pleasant and friendly atmosphere contrasted with the ferocity of the competition. The rules were strict and rigorously enforced. All participants were weighed beforehand and the differences in their weights neutralized with ballast. Removable parts were marked with "secret paint" so that they could not be surreptitiously replaced en route. Gasoline was measured to the last drop and filler caps sealed to ensure the results of fuel consumption tests were precise and reliable. In the race organizers' office, the results were continually updated on a large board. Josef Ganz, thoroughly impressed, enjoyed all this immensely. As ever, he was there as a journalist for *Motor-Kritik*. Armed with his camera, he followed the little cars in his Tatra and was surprised by some of the steep roads and poor surfaces along the route, where even the Tatra struggled. There was an atmosphere of cheerful exuberance. Everyone lived and breathed cars and motorcycles. They ate together, talked, laughed, and debated. Children waved from the roadside, smiling happily at the cars as they passed.

Ganz felt that this fierce but amiable contest would "without doubt constitute the true beginnings of the German Volkswagen." Factory driver Otto Wegmer notched up twenty points in the Standard Superior and won a gold medal. A competitor challenged Wegmer to corner the car on a gravel path at over thirty miles per hour, first in wide arcs, then tighter and tighter, until the thin tires could no longer cope and centrifugal force flipped the car over. There the little Superior lay on its back like a dead beetle, four wheels in the air. Fortunately, Wegmer was unhurt. Bystanders volunteered to help roll the car back onto its tires, and Wegmer and Ganz surveyed the damage: a few scratches and some abrasions to the artificial leather at the edge of the roof, but nothing to speak of. Wegmer got back in and drove away as if nothing had happened, leaving a small puddle of oil behind as the only evidence. The Standard Superior had come through all its tests at the small-vehicle competition with flying colors, which made Ganz more convinced than ever that this car would serve as the basis for the German Volkswagen. "The German people need transportation they can afford," he wrote in *Motor-Kritik*. He believed the government should give the designers of small cars a helping hand: "What about state aid for the creation of the German Volkswagen at a design and research institute?"[12] An initiative of a similar kind had been undertaken for the development of the Volksradio, a simple but effective mass-produced radio receiver introduced at the Berlin Radio and Television Exhibition in August 1933 for the relatively low price of seventy-six Reichsmark. There was nothing frivolous about the introduction of a People's Radio; it was a carefully considered initiative aimed at enabling Hitler's speeches and Nazi propaganda to penetrate every German home.

A Small Phenomenon

Despite his many personal problems of the past few months, Ganz was enjoying a wonderful summer. The Standard Superior, based on his patents, achieved major successes in the 2000 KM Durch Deutschland, the 1000-km-Fahrt, and the small-vehicle competition. At the Standard Fahrzeugfabrik and in the editorial offices of *Motor-Kritik*, one letter after another arrived from customers who were extremely satisfied with the little car: "Far better than I expected," they wrote. "A small phenomenon!" and "the car offers all the advantages that can make it in the most literal sense of the word a Volkswagen." A man in Dresden had driven no fewer than 10,439 miles in his Standard Superior in just three months, spending a total of just 17.70 Reichsmark on repairs. He declared: "You really couldn't expect better than that from this magnificent little car."[13]

In that summer of 1933, one Felix Korn from Ulm made a round trip of almost 750 miles through the mountains of Switzerland and Italy in the Standard Superior, and he wrote to *Motor-Kritik* that it "exceeded my expectations by a long way." Everywhere he went, people gathered around. With its little 0.5 liter two-stroke engine, the car could easily do fifty miles an hour, and Korn had reached almost sixty with the windows closed. He even overtook larger vehicles with eight-cylinder engines. It had performed impressively in the mountains too: "The acceleration was outstanding even on steep roads. Not once was the 'Superior' overtaken." He concluded: "Its reliability and efficiency [...] have been proven as far as I'm concerned by this tour of the Alps."[14] Ganz regularly received personal missives. H. I. Hoffmann, a former dealer for Studebaker in Munich, expressed admiration for the fact that with *Motor-Kritik* Ganz had helped "the German auto industry back onto its feet so quickly." "It really is possible to buy a German car again now, and we definitely have you to thank for that."[15]

The overwhelming success of the Standard Superior persuaded the Standard Fahrzeugfabrik to begin developing a new, rather larger version in the summer of 1933, designed to accommodate a family of four. The old model was extended by 250 millimeters (9.84 inches), to a length of 3.3 meters (10 feet, 10 inches). The extra space meant there was room for a small back seat directly above the engine for children or additional luggage. This version had different, more modern and appealing bodywork, with extra side windows and a small rear windscreen near the air intake for the engine. In late September, the new model was ready. The car was on offer in either silver-gray with blue-gray fenders or dark red with black fenders. On the front of the new brochure was a drawing of a family with two children proudly standing next to their Standard Superior and proclaiming: "Room enough for the four of us in the fastest and cheapest German Volkswagen." The new Standard Superior for the 1934 model year was extremely well received by the press. The magazine *Motor und Sport* wrote:

New model Standard Superior, 1933

> The new streamlined bodywork for 1934 features a substantially improved finish on the inside, which together with the harmonious color scheme and the new contours make an excellent impression. Even for

10
14

Die Autobahn

und „Die Reichsautobahn"
DAS FACHBLATT FÜR DAS AUTOBAHNWESEN

FERNSPRECHER: A 1 JÄGER 0010 · NEBENSTELLE 272
DRAHTANSCHRIFT: GEZUVOR · BANKKONTO: COMMERZ- UND
PRIVATBANK, ZENTRALE BERLIN · POSTSCHECKKONTO: BERLIN 5670

BERLIN W8
LEIPZIGER STR. 3

Herrn
Dipl.Ing.G a n z

Frankfurt/Main
-.-.-.-.-.-.-.-
Blücherstrasse 2o
(Motor - Kritik)

K/S. 21.11.33.

Sehr geehrter Herr G a n z !

Seit langen Jahren haben Sie es sich in anerkennenswerter Weise zur Aufgabe gemacht , die deutsche Autoindustrie mit neuen technischen Gedanken zu befruchten .

Die Ausführung der Reichsautobahnen wird und muss ebenfalls dazu beitragen , die Autotechnik vor neue Aufgaben zu stellen . Wir haben diese Frage in den letzten drei Heften unserer Zeitung , von denen wir Ihnen nochmals je 1 Exemplar zustellen , von verschiedenen Seiten behandeln lassen . Wir sind nun inzwischen an die Autoindustrie herangetreten , damit sie zu den angeschnittenen Fragen Stellung nimmt . Den Erfolg dieser Rundfrage werden wir in unserem Januar-Heft , welches noch vor Weihnachten in Druck gegeben wird , veröffentlichen .

Wir möchten Sie nun bitten , uns ebenfalls hierzu einen Aufsatz zu schreiben . Dies wird Ihnen umsoweniger schwer fallen , weil Sie ja nur einen Niederschlag Ihrer Gedankenwelt in dieser Angelegenheit wiederzugeben brauchen .

Für Ihre freundliche Stellungnahme zu den Arbeiten der "Hafraba" und unserer Schriftleitung sagen wir Ihnen besten Dank . Der heutige Brief wird wohl endlich einmal zu dem führen , was zwischen Ihnen und unseren Herrn K a f t a n zu Beginn dieses Jahres in einer persönlichen Unterhaltung besprochen worden war.

Vielleicht lassen Sie uns sofort wissen , wie Sie sich zu dieser Frage stellen. Das Manuskript selbst müsste bis zum 9.Dezember hier eintreffen .

Wenn Sie uns noch einige Bilder dazu zur Verfügung stellen könnten, dann wäre Ihre Mitarbeit in jeder Weise vollkommen .

" Heil Hitler !
" Die Autobahn "
Schriftleitung

Letter from the magazine *Die Autobahn und "Die Reichsautobahn"* to Josef Ganz, 1933

> large people wearing winter coats, the roomy and light interior of the car, which has four windows, offers plenty of room; there is also space in the back for two children or for large pieces of luggage.[16]

Streamlined Car for the Autobahn

Endurance tests had already shown that the Standard Superior could be driven for many hours without problems. The next step was to give it fully streamlined bodywork, so that it could achieve greater speeds despite its small engine capacity. This was intended to be the Volkswagen in which German families would soon be traveling a new highway network that crisscrossed the entire country. On September 23, 1933, Adolf Hitler, watched by his retinue of senior Nazis, personally thrust a spade into the ground to mark the building of the first stretch of Autobahn between Frankfurt and Darmstadt. With that spade, Hitler also metaphorically buried the true initiators of the highway project, a consortium of German roadbuilders that had devoted itself since 1925 to the building of a highway all the way from Hamburg in northern Germany to Genua in northwestern Italy. They called themselves the Association for the Building of a Road for Rapid Motor Vehicle Traffic from Hamburg via Frankfurt am Main to Basel, and their initiative was generally referred to as the HaFraBa project. Hitler's decisive approach had transformed their dream into the official Autobahn of the glorious Third Reich. Now for the Volkswagen. A photograph of Hitler with his spade, headlined "Start of Construction of the Reichsautobahnen," was published on the cover of the issue of *Motor-Kritik* in which Josef Ganz presented the new styling of the Standard Superior to his readers.[17]

Ganz was extremely enthusiastic about the building of the new highway network. He had been a supporter of the original HaFraBa project for years and enjoyed a good relationship with the editors of the specialist magazine on the subject: *Die Autobahn und "Die Reichsautobahn"— das Fachblatt für das Autobahnwesen*. Earlier in 1933, he had discussed with its editor-in-chief the need for a radically different type of car that would allow people to travel efficiently at high speeds on the new roads. A streamlined car with a rear-mounted engine was ideal for the purpose, and its development had received an enormous boost now that the starting gun had finally been fired for the roads project. The editors of *Die Autobahn und "Die Reichsautobahn"* sent Josef Ganz a special invitation to write an article on the subject for their magazine. Their letter opened with the words: "For many years you have made it your task, in a commendable fashion, to stimulate the German auto industry with new technical insights."[18]

More and more engineers were realizing that the streamlined car would be ideal for the new highways. *Die Autobahn und "Die Reichsautobahn"* published articles in quick succession by Josef Ganz and his archenemy Paul Ehrhardt in which they sang the praises of streamlining for passenger vehicles. In his article, Ganz concluded there could be no doubt that "the building of highways" would give "a powerful boost to the technical development of the automobile."[19] The streamlined car became a topic of conversation throughout the German car industry and the German media. The country was captivated by the idea. "One can do it differently but—only streamlining will help!" the motoring magazine *Automobilia* declared in September 1933 on the subject of the ideal highway car. "Not, of course, the streamlining that is now being offered by fashionable bodywork designers as the latest fad, but solely the scientifically researched aerodynamic shape, as developed in Germany by Jaray and Rumpler, which Ganz has championed in the literary sphere."[20] In imitation of the Standard Superior, more and more manufacturers were beginning to experiment with rear-mounted engines and streamlining.

Innovation in Paris

In October 1933, as he did every year, Josef Ganz traveled to Paris with his colleague Walter Ostwald for the Salon de l'Automobile. On the day before the opening of the show, walking along the Champs-Élysées past the Grand Palais with Ostwald, he watched the bizarre process of getting the cars into the exhibition complex, which dated from 1900. One by one they were loaded onto huge wooden carts with solid tires before being pushed backwards by teams of horses toward the raised entrance. The following day, images of the opposite extreme in vehicular transportation were on show. Chenard & Walcker, for example, was exhibiting its extravagant streamlined car with bodywork by the French coachbuilder Mistral, which could reach speeds of hundred miles per hour, and Bugatti had brought its flamboyant Type 49. Several manufacturers had seized upon streamlining as a fashionable term, including Peugeot, which was advertising independent front suspension and aerodynamic bodywork as features of its models for 1934.

Among those Ganz and Ostwald spoke to at the show were the director André Citroën, who was developing a new front-wheel-drive model with independent suspension, and engineer Joseph Dauben, now working under Gustav Röhr at Adler. Although Dauben had developed successful front-wheel-drive cars at both Röhr and Adler—the Adler Trumpf was being built on license by Rosengart in France—he too expressed the opinion that developments within the auto industry were pointing in the direction of the streamlined, rear-engined car.[21] Josef Ganz had great respect for Dauben, who in his view had developed some extremely progressive models. As a result of this meeting, editor-in-chief Georg Ising published an article in *Motor-Kritik* in which he presented Dauben as a "hidden talent," the spiritual father of all the designs produced by Röhr and Adler, for which Gustav Röhr had accepted all the acclaim. "It is to Röhr's great credit that he has brought this man to the fore," Ising wrote. "Röhr as manager, Dauben as artist and talent."[22] Gustav Röhr was furious when he read the article. Soon after the Paris show was over, he went to Reichsstadthalter Jakob Sprenger again and handed him a pamphlet written by Paul Ehrhardt with the title "*Motor-Kritik*, the Sensationalist Rag of the German Auto Industry."[23] The criminal investigation into Josef Ganz was still underway. If it were up to Ehrhardt, Röhr, and their accomplice Bernhard Stoewer, short work would be made of Josef Ganz and his "sensationalist rag."

CHAPTER TWELVE

HITLER SKETCHES THE BEETLE

In the summer and fall of 1933, Jakob Werlin had countless formal and informal discussions with the directors of manufacturing firms such as BMW and Daimler-Benz about the development of a Volkswagen. One after another they told him it was impossible to build a People's Car for less than a thousand Reichsmark—the maximum advocated by Josef Ganz in *Motor-Kritik* and the price to which Hitler had become so attached. The only interest in the project shown by individual manufacturers was in promoting their own existing and forthcoming models as Volkswagens. Opel tried to persuade Hitler to support it in building the cheapest mass-produced "Volkswagen," a conventional model costing RM 1,880 that could be made even cheaper if produced in larger quantities.[1] Opel was already advertising the car as the "German Volkswagen," and its promotional campaign was supported by some sections of the press,[2] yet it bore no resemblance to what Hitler envisaged a Volkswagen to be. He wanted a technically progressive model, a scientific tour de force of which Germany could be proud—a car designed, in fact, along the lines of the Standard Superior that had proven its worth over the past six months.

Independent Engineer

Jakob Werlin had come up against the same brick wall as Josef Ganz: the auto industry. It simply did not want to have to compete with a cheap, technologically advanced Volkswagen, nor was it willing to make its engineers available for the development of such a car by a state consortium. This was the crux of the matter. Almost all Germany's competent engineers were employed by car manufacturers. Only a handful of suitable independent engineers remained, including Ferdinand Porsche, Josef Ganz, and Edmund Rumpler.

Of the three, it was Ganz who had championed the Volkswagen, explaining in *Motor-Kritik* over the years what the main ingredients would be: a technical design on the Maikäfer model with a streamlined "bug shape," a maximum retail price of a thousand Reichsmark, promotion of the name Volkswagen, and a scheme to combine the knowledge of the best engineers from all across the German car industry in a single state-sponsored project. Ganz was the driving force behind the recent wave of innovation in the German auto industry, and he had been present at the birth of progressive models such as the Mercedes-Benz 170, the BMW AM1, and the Standard Superior. Edmund Rumpler was another man with a creative spirit. He had decades of experience, having produced innovative aircraft as well as vehicles like the Tropfenwagen and a streamlined truck with front-wheel drive and swing axles. Ferdinand Porsche, of the same generation as Rumpler, had thirty-five years of experience in auto technology but had turned his attention to small models based on the latest ideas only within the past two years. The first such cars for which he was responsible were the Zündapp Type 12 prototypes, followed in late 1933 by three similar prototypes for the car and motorcycle manufacturer NSU. The NSU Type 32 had independent suspension with swing axles and a rear-mounted, air-cooled, four-cylinder boxer engine. Of the three prototypes, two had steel bodywork, while the other was made of wood covered in artificial leather with metal fenders, a structure similar to that of the Standard Superior. The project had been halted when NSU entered into a license agreement with Fiat in Italy that meant it could no longer develop its own models. In late 1933, Porsche's design bureau was continuing to work on the state-sponsored Auto Union racing car. The first prototype was tested that November on the Nürburgring.

Of the three engineers, Josef Ganz and Edmund Rumpler were soon out of the running, not because of any lack of technical ability, but because both were Jewish. Like Ganz, Rumpler had been arrested and imprisoned by the Gestapo, but he was set free after intervention by Hermann Goering, who greatly respected Rumpler for his pioneering aviation work before and during the First World War. After having attended a Nazi rally, Rumpler's wife Julie felt revolted and physically sick by Hitler and decided immediately to get her family out of Germany. Their three daughters and one son were sent abroad to Austria, England, Italy, and Switzerland. The only one she could not persuade to leave was her husband, Edmund. He initially did not believe that Hitler was the evil that his wife instinctively knew he was. Rumpler remained in Berlin, but soon he was not allowed to work and could not own a telephone. All of his property and his money were confiscated, but on the order of Goering, he was given a monthly allowance by the government to live on.[3]

In practice, Werlin had little choice, although he did value Porsche as an engineer, with his "theoretical knowledge," "an excellent perspective on practical matters," and "the gift of finding the right staff and deploying them appropriately." The two men had known each other for many years, both having been racing drivers in Austria before the First World War, Porsche for Daimler-Benz and Werlin for Puch. Werlin first discussed the option of appointing Porsche to head a consortium to develop the Volkswagen with his own current employer, Daimler-Benz, but its technical manager and former racing driver Max Sailer wanted none of it, and its director Wilhelm Kissel had a wait-and-see attitude.[4] At that point, Josef Ganz was still a technical consultant for Daimler-Benz, and he had been of immense value to the company during the development of the Mercedes-Benz 170 and the 120H prototype. Now Daimler-Benz was about to introduce the final result, the Mercedes-Benz 130H, modeled on the Maikäfer, with a rear-mounted engine and a backbone chassis to whose central tube were attached the independently suspended front wheels, with their leaf springs and steering mechanism, and the independently suspended rear wheels with coil springs and swing axles.

Jakob Werlin had been neglecting his responsibilities as a salesman for Daimler-Benz to such an extent that the company wanted to sack him. In fact, an accountant had been hired to discharge him from his position. But to give Werlin more influence as he went about his difficult task of developing the Volkswagen, Hitler intervened and had him promoted onto the board of Daimler-Benz in October 1933. The plan was now to make Ferdinand Porsche head of a developmental consortium for a National Socialist Volkswagen project, and for prestige reasons it was of crucial importance for the Nazi government to obscure the huge contribution Josef Ganz had made to the car's development. This necessitated the termination of his contract with Daimler-Benz, a firm that was to be an important partner in the creation of the German Volkswagen. Only a few weeks after Jakob Werlin's appointment to the board, Josef Ganz in Frankfurt received a letter from Daimler-Benz. He immediately tore open the envelope and read:

The chassis of the Mercedes-Benz 130H, 1933

> Circumstances have developed such that it is no longer possible for you to provide your services to us in the way we had in mind when we drew up the contract. We therefore regard it as in our mutual interest to arrange for an early dissolution of the current agreements between us as regards technical advice by you and the placing of your inventions at our disposal. We accordingly terminate both agreements with immediate effect.[5]

On reading the brief, businesslike letter, Josef Ganz had a bitter feeling that after an extremely fruitful period of cooperation with Daimler-Benz lasting nearly three years he had suddenly been pushed aside. The terms of his second contract stated that it could not be terminated before December 31, 1935, so he appealed against the decision with the help of a Berlin lawyer.[6] This meant he still had to ask permission from Daimler-Benz every time he wanted to sell rights to designs he had patented. That was the situation in November 1933 when he was approached by Everhard Bungartz, director of the recently founded Bungartz & Co., who had been greatly impressed by the Standard Superior and wanted to put into production a small car based on Ganz's patents. It would be called the Butz, and it was technically almost identical to the Standard Superior. Daimler-Benz gave permission for a license to be granted to Bungartz on condition that the car remained within the 250-400 kilogram class (551-882 pounds).[7] This would prevent the Bungartz Butz from competing with the Mercedes-Benz 130H.

Bungartz Butz Cabrio-Limousine, 1934

Tatra Files Charges against Ganz

As well as Bungartz, Tatra had now started to develop new models based on Maikäfer principles. Ganz heard through one of his contacts in the industry that a mysterious five-and-a-half-yard-long streamlined prototype from Tatra with a rear-mounted, air-cooled, eight-cylinder, three-liter engine was reaching speeds of well over ninety miles per hour, with excellent road holding. Moreover, Tatra was looking at the "issue of a small rear-engined car."[8] Tatra would present its large streamlined model, the Type 77, at the upcoming IAMA in Berlin. A second, much smaller model featured a rear-mounted, air-cooled, two-cylinder engine, and only one prototype had been built. This car, the Tatra V570, closely resembled the Standard Superior both technically and in appearance. It was Tatra's attempt to put the ultimate Volkswagen on the market. The company was also pursuing efforts begun earlier with the help of Paul Ehrhardt to claim intellectual ownership of all cars featuring a backbone chassis, swing axles, and a rear-mounted engine. When Ehrhardt presented the Gestapo with false charges against Ganz in April 1933, it had been partly with the aim of gaining possession, on Tatra's behalf, of his Volkswagen design and patents.

One ray of light in those dark winter days was that in December the public prosecutor at the court in Berlin finally dismissed all the charges against Ganz based on accusations by Ehrhardt, including blackmail of the auto industry.[9] Strict censorship by the Nazi Party meant that *Motor-Kritik* could publish only a simple statement, without comment: "Investigation proceedings against Dipl.-Ing. Ganz (see *Motor-Kritik* no. 11 & 12, 1933) ended."[10] This did not mean the case was closed. Ganz was determined to drag Paul Ehrhardt, Gustav Röhr, and Bernhard Stoewer, the men behind his arrest, before the courts. He was playing with fire. It was an attempt to obtain justice in a constitutional state that was rapidly being transformed into a totalitarian dictatorship in which Jewish citizens had lost all their rights. Meanwhile, Tatra continued its attacks on him with as much ferocity as ever, wrongly accusing him in mid-December of violating its intellectual property rights as defined in patent DRP 549602[11] with his patent DRP 587409.[12] Insofar as there was any similarity between the two patents, it lay in the fact that they both described the positioning of a mid-mounted engine, but the rest of the car was quite different in the two cases. The Tatra patent described a three-wheeled car with a drive shaft from the engine to the single rear wheel, whereas Ganz's patent was for an engine block mounted in a different way, with direct transmission to swinging rear half-axles.

Just recently Austrian engineer Béla Barényí had asked Josef Ganz if he was interested in a joint application for his German patents. Like Ganz, Barényí had made designs for a Volkswagen but lacked financial means for international patent applications, since he had been without regular work. Ganz promised Barényí to look into the matter—unaware that such ventures would soon become impossible—and also told Barényí that he had observed a "great talent for drawing" in his designs which "could be put to use" at *Motor-Kritik*.[13] One of Barényí's streamlined car designs would soon after appear on *Motor-Kritik*'s front cover.

Hitler's Sketch

As Christmas approached, Josef Ganz stayed at home in Frankfurt. It was a strange and disturbing time. Within the space of six months, he had been dismissed as technical consultant to BMW and Daimler-Benz and lost his job as editor-in-chief of *Motor-Kritik*. Those three contracts had brought him an annual income of almost thirty thousand Reichsmark. As a result of his many problems of the past year, he had started to suffer from anxiety and from attacks of migraine; Madeleine Paqué was extremely worried about him. When not resting in bed or on the sofa, he withdrew to his editorial office to work on the next issue of *Motor-Kritik*. It was a great frustration to him that as the founder and editor-in-chief of the influential trade magazine he had been pushed aside on the orders of the Nazi Party. Despite the severe personal setbacks of the past few months, he was at least still reasonably hopeful about the future of the Volkswagen. The new license for Bungartz meant he was busy designing and sketching streamlined bodywork for the car he had in mind. In that, he was not alone.

In his house in Berchtesgaden on the Obersaltzberg in Bavaria, Chancellor Adolf Hitler was also sketching designs for small cars. Since the summer, Hitler had been determined to build an uncompromising Volkswagen of the kind for which *Motor-Kritik* was continuing to

Sketch by Adolf Hitler, December 23, 1933

publish powerful propaganda. He intended it to be the innovative German counterpart to the Model T Ford that had brought such success years earlier to his anti-Semitic idol. In the United States, one in five people owned a car; in Germany, one in 140. In those quiet, dark days before Christmas, Hitler daydreamed about what *his* Volkswagen should look like, creating the emphatic, sharply penciled outline of his first rough sketch. Faithful to the design guidelines provided by *Motor-Kritik*, Hitler depicted a thoroughly streamlined car with a short nose, a long, flowing, descending roofline, and a panoramic front windshield. It had two doors halfway along the bodywork and rear wheels placed far back, presumably to make space for the mid-mounted engine advocated by Josef Ganz, placed toward the back of the vehicle but in front of the rear axle. Hitler signed his sketch with the initials A. H. and the date December 23, 1933.[14] It would be the ideal car for driving on the highway network for which he had symbolically plunged a spade into the ground in September. In late December, he showed the sketch to Jakob Werlin and asked him how his negotiations with the industry were going. To Hitler's frustration, there was little to report. All hope now lay with Ferdinand Porsche. Shortly after his meeting with Hitler, Werlin decided to call in at Porsche's office in Stuttgart without prior warning to gauge his interest.[15] Werlin hinted to Porsche that there were serious plans for a state-financed Volkswagen project and asked him to put his vision of the People's Car on paper and deliver a copy without delay to the Ministry of Propaganda. Werlin left the office in Stuttgart feeling optimistic. As soon as he was gone, Porsche started writing a proposal in which he explained to the government his plans for the development of the "German People's Car."

Mercedes Test Drive

Despite the termination of his contract with the manufacturer, as a journalist for *Motor-Kritik* Josef Ganz had been allowed to borrow a brand-new Mercedes-Benz 130H Cabrio-Limousine from the Daimler-Benz factory immediately after Christmas and had driven it back to Frankfurt. At home, outside the door of the villa on the Zeppelinallee, he photographed the car from all sides before setting off with *Motor-Kritik*'s editors Georg Ising and Walter Ostwald for an extensive test drive through the snowy Black Forest. The 130H represented the next phase in Daimler-Benz's innovation campaign. Although the car was based on the Maikäfer concept, one important difference was that the engine, against Ganz's advice, had been mounted upright and behind the rear axle. At the front of the car were the fuel tank, the spare wheel, and a small space for luggage and tools. "The MB 130 isn't a proper rear-engined car," Ganz wrote in his evaluation of the new model in the next issue of *Motor-Kritik*. "It's more an automobile with an 'outboard motor.' The engine has been built out onto the back, which gives it an undesirably troublesome tail."[16] The effect was exacerbated by the deep snow on the roads of the Black Forest.

Hope for the New Year

The day before New Year's Eve, Ganz received a second letter from the publishers of *Motor-Kritik*, confirming the termination of his contract.[17] To its great regret, at the time of writing the publishing company, with which he still had a good relationship, could do nothing more than to "wish him a good 1934." Doubts about what that year would bring were increasing by the day. In their introductory editorial for the New Year's issue of *Motor-Kritik*, the editors wrote that they were looking forward to 1934, since the past twelve months had been "full of important events for the German auto industry." They also expressed the hope that they would now see an end to the "continual plotting against *Motor-Kritik*" and that in an "unsullied atmosphere" it would be possible to work cheerfully on the further "reconstruction" of the German auto industry.[18] With his recent dismissal by Daimler-Benz in mind, Ganz used the introduction of the Mercedes-Benz 130H as an opportunity to tell his story in the magazine. Although he could not go into matters such as his arrest by the Gestapo and his forced dismissal as editor-in-chief of *Motor-Kritik*, he could tell his faithful readers one last time that he had been present at the birth of the Maikäfer concept, which was now starting to break through on a large scale. His article read like a technical autobiography. Ganz described how his idea of designing a Volkswagen had arisen more than ten years earlier and told of the positive test results with the Benz Tropfenwagen in 1930. He went on to describe the development of the Ardie-Ganz and Adler Maikäfer prototypes and the production models that flowed from them: the Standard Superior and the Mercedes-Benz 130H.[19]

American New Year's greeting to *Motor-Kritik*, 1934

While many thousands of copies of *Motor-Kritik* were being distributed at home and abroad, Ganz spent the beginning of the new year preparing his case against Paul Ehrhardt, Gustav Röhr, and Bernhard Stoewer. To add to the evidence he had been collecting for two years, in early January he went with Georg Ising to visit director Ernst Decker of the Neuen Röhr-Werke in Ober-Ramstadt. During a long conversation, Decker, a former employee of Gustav Röhr, told them in detail how Röhr had purloined several items of company property after his factory went bankrupt. In total, sixty-four out of a larger number of stolen technical drawings had been seized during a search of Röhr's house. An engine prototype was found in the back of a car in Darmstadt that was owned by an associate of Röhr. Still missing was a valuable prototype Röhr had taken from the factory by distracting the guards while it was driven out through the factory gate by an accomplice.[20]

In early February, Ganz and Ising paid another joint visit, this time to Ludwig Theilen, the owner of a repair shop in Darmstadt. Theilen was able to tell them how he and his brother had been commissioned by Röhr shortly after the last war to build cars out of parts that Röhr had bought up from stocks held by the German army depot. Röhr had gotten hold of the many additional parts that were needed to finish the job from the warehouse of the car manufacturer Priamus, of which he was technical director, without paying for them. The cost to Priamus had added up to an astonishing RM 150,000, which bankrupted the firm. Röhr made an enormous profit on the sale of the cars as a result of the hyperinflation of the early 1920s, since he refused to adjust the amount he paid Theilen and his brother to take account of massive price increases. Theilen had ended their cooperation at that point, but he nevertheless became caught up in a complex police investigation into the misappropriation of an aero engine that the engineer Joseph Dauben had developed for Röhr during the war with a subsidy of a million marks. Two finished engines had been hidden from the British occupying forces by an acquaintance of Theilen's brother. In an attempt to protect his brother from prosecution, Theilen had himself become involved and spent two weeks in custody. The case was finally dropped when no trace of the two engines could be found. At the last moment, they had been destroyed and the parts thrown into the Rhine. Theilen was freed on bail of RM 5,000, put up by his brother, a sum that despite frequent promises Röhr had never paid back. Theilen described Röhr as a criminal and expressed serious fears for his own personal safety should he ever make the story public and seek restitution from Röhr and his accomplices. He believed Röhr was aware of the case Ganz intended to bring against him. Several weeks before Ganz and Ising visited Theilen, Röhr had sent someone to offer him a position at Adler that would exist only on paper, with the aim of making Theilen dependent and thereby silencing him.[21] The evidence against Paul Ehrhardt, Gustav Röhr, and Bernhard Stoewer was piling up.

Porsche's Exposé

That January, while Ganz was working on his court case, a copy of *Exposé betreffend den Bau eines deutschen Volkswagens* (*Exposé Concerning the Building of a German Volkswagen*) by Ferdinand Porsche arrived at the Ministry of Propaganda. In his "exposé," Porsche wrote that the Volkswagen should not be a small car at all but an all-purpose vehicle of normal size and performance, which could be mass-produced cheaply because of its low weight. He proposed a car with a track width of 1,200 millimeters (just over 47 inches), a wheelbase of 2,500 millimeters (roughly 98½ inches), and a weight of 650 kilograms (1,433 pounds). The car would have all-round independent suspension with swing axles at the rear, a maximum speed of over sixty miles per hour, and fuel consumption of almost thirty miles to the gallon. The engine, mounted behind the rear axle, would need to produce twenty-six horsepower. He suggested there should be two types, one with an air-cooled, four-cylinder boxer engine and the other with a two-stroke radial engine with three cylinders. The bodywork would have "the ideal streamlined shape." The retail price of this German Volkswagen would be RM 1,550, putting it into the same price bracket as the Standard Superior. Porsche concluded:

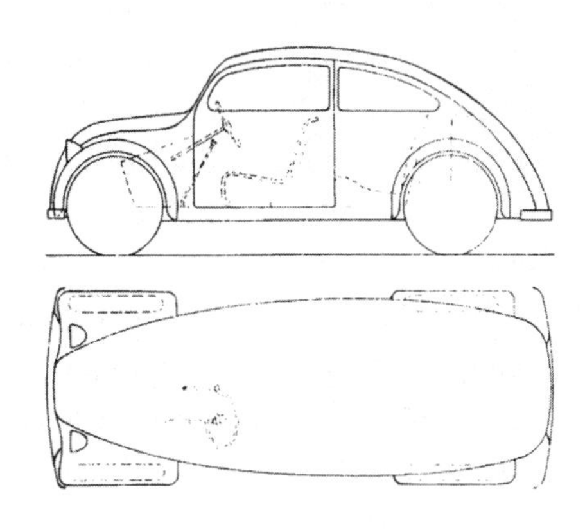

Drawing of the Volkswagen in Porsche's exposé, 1934

> The government can entrust to me the building of a Volkswagen as a research subject. The car will be fully developed and extensively tested by my office within about one year, after which a definitive test will be carried out by a commission composed of official and private experts, with the involvement of the industry. If the test results are satisfactory, the government will be able to decide to recommend to the industry the serial production of this model as the German Volkswagen. [...] The building of a German Volkswagen, for which there is expected to be a huge demand in the future among a broad spectrum of the population,

> suggests itself to me very strongly. I therefore have every reason to predict that given the situation as it stands, the government will assign to me the task of building the German Volkswagen.[22]

The editors of *Motor-Kritik* still knew nothing about the possible involvement of Porsche in a state-financed Volkswagen project of the kind Ganz had been advocating for so long. At the end of January, the first snippets of news reached them of test drives made by Hans Stuck on the AVUS in an Auto Union racing car developed by Porsche's design bureau and heavily subsidized by Hitler, under the supervision of the racing team's leader, Willy Walb. In *Motor-Kritik*, Georg Ising described how journalists were throwing themselves upon the breathtaking new racing car like hungry wolves, and one magazine after another was writing triumphantly: "Reshaped according to the German way of building a chassis, with ideal suspension and road holding. Fabulous progress! What we have always advocated has become a reality!" Ising compared *Motor-Kritik* to "a chef who looks into a packed room where many hungry guests are enthusiastically enjoying the dishes placed in front of them, which they find delectable. So delectable that not one of them any longer remembers that until recently he claimed he found such fare inedible." He said it was a "victory" for *Motor-Kritik* that "chefs" like Porsche were now among those able to "garnish" the *Motor-Kritik* recipes extraordinarily well. He closed his article with the words:

> The mastery of Dr. Porsche and the experiences of Walb and Rosenberger have materialized in the new Auto Union racing car, which is now exuberantly surging out into the world. An inconspicuous May Bug from Frankfurt once used to hum a little tune about road holding and rear-mounted engines.[23]

That other "*Motor-Kritik* recipe," the Volkswagen, was now being taken very seriously indeed. Within government circles, Hitler had openly expressed his support for the development of a People's Car. On February 12, 1934, a meeting was convened with representatives of the Ministry of

Propaganda, the Ministry of Economic Affairs, and the automobile trade association the RDA to look at ways of assessing the "economic benefits" of a Volkswagen costing less than a thousand Reichsmark and to discuss whether the government should provide a stimulus by contracting an "outstanding designer."[24] A week later, the RDA put its opposition to the development of an official People's Car in writing, asking the Minister of Transportation to hand the task over to the industry, "which has been engaged with the problem for years."[25] Hitler's will, however, was far stronger than the hedging and whining of car builders. In February, while the RDA was still trying to block Hitler's plans, the chancellor was in Berlin working on his opening speech for the upcoming IAMA motor show, in which he would publicly express his support for the Volkswagen.

Well over three hundred miles away, in Frankfurt, Ganz was working on his preview of the IAMA for *Motor-Kritik*. Three new models bearing his signature would be unveiled in Berlin—the redesigned Standard Superior, the Bungartz Butz, and the Mercedes-Benz 130H—not to mention a wide range of new models from other manufacturers that showed signs of his influence. Ganz expressed his concern, however, that the German auto industry's "fleet," which as a result of the "carelessness" and "failures" of its "captains" had found itself in "terrible straits" but had managed to get through the most dangerous period under the guidance of "skilful pilots," was now throwing those pilots "overboard," convinced it could manage alone from this point on. It would be fatal to indulge such "human weaknesses," Ganz said. "Its technical lead has not produced a final victory for the German auto industry." You only had to look at the "fate of the streamlined, rear-engined car." He went on:

> It goes without saying that precisely those most keen to boast of being pioneers could at the very least be accused of having done nothing at the time to prevent things ending up on the scrapheap that in due course could have given the German auto industry an unbeatable lead in the world market. The notion that the influence of these "pioneers" was too weak to prevent technical barbarism is not credible. Either they did not realize the significance of the new ideas, or they lacked the

> leadership qualities to persist in what they knew to be right, staking their own positions if need be. This wailing about how building methods based on the latest technical insights, which they, after all, put onto the market, have not been prevented from spreading out across the world does not say much for their competence. If they had been leaders, then their names would now be connected with designs created under their leadership over the past few years and comparable to those already common in America, but so far ahead that they could proudly point to an era they had left behind them, one that in other countries has only just begun.[26]

Ganz was referring to work such as that of John Tjaarda, a Dutchman who had emigrated to the United States in the early 1920s, where he designed streamlined cars with mid-mounted engines called the Sterkenburg series. In 1933, the Briggs Manufacturing Company in Detroit had built a full-sized mockup that was featured in *Motor-Kritik,* followed in 1934 by a running prototype. It was not destined to go into production, but the bodywork design would reappear as the front-engined 1935 Lincoln Zephyr, the first successful streamlined American car.

Although the article by Josef Ganz was published several days before the official opening of the IAMA in Berlin, Jakob Werlin would use it to try to prove that he had publicly criticized Hitler's glorious opening speech at the show.

CHAPTER THIRTEEN

KAISERHOF HOTEL, BERLIN

The Kaiserdamm in Berlin had never been so busy as in those early March days of 1934. Around the clock, trucks came and went, delivering the latest models of cars and motorcycles, along with wooden packing cases containing powerful engines, to the exhibition complex. Carpenters, electricians, and mechanics crisscrossed the floor, building the exhibition in just a few days. This year's IAMA was to be a show to beat all previous motor shows. Josef Ganz arrived in Berlin several days before the official opening to watch the preparations. In just a few days, two adjoining exhibition halls were filled with the products of an industry that had seen unprecedented progress over the past year. It struck Ganz that the German manufacturers were all showing sound and purposefully designed cars, rather than the "ephemeral models" of years past. Everywhere, new types could be seen sporting lightweight backbone chassis, independent suspension, swing axles, and the latest streamlined bodywork. Several even had rear-mounted engines. As well as the Standard Superior comparably small models were being launched, such as the Bungartz Butz, the Hansa 400, and the Framo Piccolo.

Mercedes-Benz was introducing its midrange model 130H, and on the Tatra stand the mysterious, luxury, streamlined Type 77 was covered by a large white sheet, waiting to be unveiled. The showpiece of National Socialism at the IAMA was the Auto Union racing car, developed by Ferdinand Porsche's design bureau and prominently displayed next to a vast eagle bearing a large swastika, to the honor and glory of the Third Reich. On March 6, 1934, the day before the official opening, the race driver Hans Stuck set no fewer than three world records in the streamlined lightweight racing car with its mid-mounted engine and swing axles. These cars were all built according to the Maikäfer principles for which Ganz had been campaigning so relentlessly over the past few years.

Hitler's Speech

For the official opening on March 7, Hitler's big luxury Mercedes-Benz convertible drew up outside on the Kaiserdamm, while inside the exhibition halls the last smudges and nicks on gleaming paintwork were polished away. He was welcomed at the door by Robert Allmers of the RDA. Like his entourage, Hitler was dressed in military uniform, complete with black boots—a menacing contrast to the simple two-piece suit of a year before. A newsreel camera rolled, press photographers clicked voraciously, and men and women in the street beamed with enthusiasm as they gave the "Heil Hitler!" salute. This IAMA had a strong political charge to it as a showcase for the success of National Socialism. The building was decorated inside and out with giant swastika banners and flags. Even the poster advertising the IAMA doubled as propaganda.

After an introductory address by Robert Allmers, it was Adolf Hitler's turn. Behind him stood dozens of SA ensign bearers, holding their outsized flags high. Surrounded by newsreel and radio microphones, Hitler spoke. He expressed his joy at the technical innovations in the auto industry over the past twelve months but said it was "a source of bitterness" that the motor car was still a means of transportation beyond the reach of millions of "good, diligent, competent people." The National Socialist government was determined that by stimulating the German car industry it would not only provide jobs for hundreds of thousands of people but put hard-working folk in a position to buy a car of their own, a Volkswagen. To illustrate how this could be achieved, Hitler pointed to the Volksradio, developed over the past year and already successfully put into mass production. He said: "I would now like to single out as the most important task of the German auto industry the building of a car that is bound to open up a market of a million new buyers." He described it as the duty of every German to declare solidarity with the millions of jobless among the German people and ended his speech with the words: "With pride and joy, I hereby declare the 1934 Berlin international motor show open!"[1] Loud applause followed. His audience was hugely enthusiastic about Hitler's highly effective performance—but the established car manufacturers were less happy with his open support for the development of a People's Car.

After his speech, Hitler toured the floor of the show. Among his entourage were Reich Marshal Hermann Göring, Minister of Propaganda Joseph Goebbels, Adolf Hühnlein of the NSKK, and Jakob Werlin of Daimler-Benz, Hitler's unofficial adviser on automotive matters and the man charged with the development of the Volkswagen. For National Socialist reasons, they first visited all the German manufacturers. At the Mercedes-Benz stand, Hitler absorbed detailed information about the latest models, including the Type 500 Kompressor, which had streamlined bodywork, and the 130H, with its rear-mounted engine. Hitler had already heard a good deal about the 130H from Jakob Werlin, and he was extremely interested in this new way of building cars. He and his entourage then cast their eyes over offerings from other German firms such as Opel, BMW, Hanomag, Adler, and DKW, in descending order of importance, until they arrived at the smaller makes, like Bungartz, Framo, Hansa, and Standard, each of whose stands featured at least one small model with a backbone chassis, independent suspension with swing axles, and a rear-mounted engine. The Standard Fahrzeugfabrik display, where the new Standard Superior was on show, along with a sports version, was flanked by tall swastika banners. Next to the Standard Superior sat a pile of brochures declaring the car to be "the fastest and cheapest German Volkswagen" for the average German family of four. Its price, as at the previous show a year earlier, was RM 1,590.

Car of the Future

After dutifully visiting all the German displays, Adolf Hitler went straight to the Tatra stand. He secretly preferred the Czechoslovakian firm's cars above all others, and at this year's IAMA it was presenting its extraordinary new streamlined model, the Tatra 77. Nearly 16½ feet long, the car was an imposing sight with its smooth, streamlined shape, its three-section panoramic front windshield, and its central fin running from halfway along the roof to the back, over the hood. The air-cooled V8 engine was mounted behind the rear axle. Separately, next to the car, stood the impressive engine block, built as a single unit with the gearbox, and attached to it the independently sprung swing axles with their rear wheels. Technical director Hans Ledwinka was present in person, ready to demonstrate the Tatra 77's special features. This was the kind of car Hitler wanted to see on his Autobahnen, precisely as described year after year in *Motor-Kritik*, looking the way he had sketched it himself in late December. He broke into a broad smile when he saw it and was genuinely fascinated by its technical details. The show was so busy that he asked Ledwinka to come to his hotel room that evening to tell him more about the new model. As always when he visited Berlin, Hitler was staying at the luxurious Kaiserhof Hotel on the Wilhelmplatz, opposite the Chancellery. In the wake of Hitler and his entourage, the general public and journalists streamed into the exhibition halls, and now Josef Ganz too could finally examine the mysterious streamlined Tatra from close proximity. He had mixed feelings about the car. The Tatra 77 was the most uncompromising application of the concept he had been promoting in *Motor-Kritik* for years, but he was less happy with the way Tatra was trying to lay claim to that concept and eliminate all competition. He had been enmeshed in a dispute with Tatra over patents for several months. This delicate subject hung in the air while he chatted with Hans Ledwinka and congratulated him on his latest creation. He took several photographs of the Tatra 77, helped himself to a folder from the pile, and walked away across the exhibition floor.

Leafing through the folder, he read: "Tatra Type 77, the forerunner of a forthcoming series of developments in automobiles. The car that turns the future into the present!"[2] On the final page he suddenly noticed the name of the author: Paul G. Ehrhardt. Ganz had not thought about it before, but at any moment he might run across his archenemy. He continued on his way rather nervously. A little later, he did indeed see Ehrhardt in the distance with his other adversary Gustav Röhr, hovering near the Adler stand. He had not seen or spoken to either man for some time, but since January he had been gathering evidence for his court case against them. That evening, as Hans Ledwinka was enthusiastically telling Adolf Hitler in the Kaiserhof Hotel about the creation of the new streamlined Tatra 77,[3] Josef Ganz was at the Auto Hotel—a modern establishment for the motorist that had its own gas station—typing out his report on the IAMA for *Motor-Kritik*. He wrote:

> Foreign countries certainly made an effort to give the impression that they had outstripped us fair and square. The folder from the Tatra factory states that to it was left the task of producing the car of the future, the streamlined automobile with its engine at the rear. To make such a claim it needed to banish all talk of its German precursors, but the fact is they cannot be beaten into submission, whether the Standard Superior or the Hansa 400, which stood on the Kaiserdamm a year ago, and the people of Stuttgart have often seen the Mercedes-Benz 130 and the Porsche prototype, which come strikingly close to the current Tatra 77 in both structure and appearance. In our view, it would have been more sensible for Tatra to have left out the claims to originality it makes in its folder.[4]

Praise for Josef Ganz

To experts at the IAMA, the identity of the spiritual father of the modern, streamlined, rear-engined car was crystal clear: Josef Ganz, now unanimously praised by the German press. *Automobiltechnische Zeitschrift* (*ATZ*) described at length the lonely battle he had fought over the years for modern technical principles, while the rest of the specialist press remained "almost entirely silent." "Meanwhile, the magazine *Klein-Motor-Sport* had acquired the title *Motor-Kritik* and published content to match. In its pages, from that point on, progress in car building found the most decisive approval."[5] The auto magazine *Motor* concluded:

> It is impossible to deny that the matter of the rear-mounted engine—powerfully advocated for years by Dipl.-Ing. Ganz—is emerging very strongly. We should think not only of the four designs for small cars by Framo, Hansa, Standard, and Butz, which are all more or less consciously or unconsciously modeled on the Standard Superior, but above all of two extremely significant innovations, the little 1.3 liter Mercedes-Benz and the eight-cylinder Tatra. Perhaps, based on these types, the chassis designs of all passenger cars will change completely.[6]

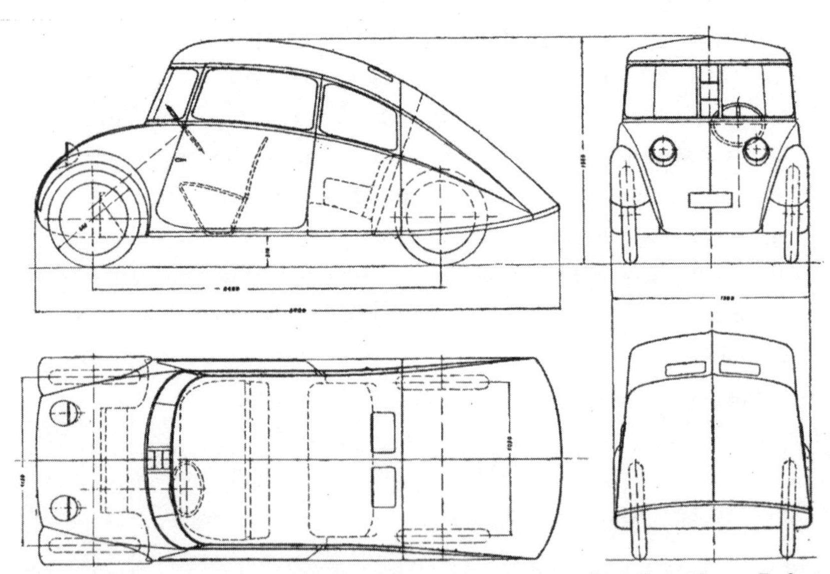

Bauvorschlag für einen viersitzigen Gebrauchswagen von Dipl.-Ing. Ganz. Radstand 2450, Spur vorn 1150, hinten 1050, Gesamtlänge 3700, größte Breite 1385, Gesamthöhe 1550, Bodenfreiheit 210 mm. Jaray-Stromlinienkörper-Vollschwingachser mit querliegendem Heckmotor.

Volkswagen design by Josef Ganz, 1934

Even Ganz's greatest rivals came round. Editor-in-chief of *Motor und Sport* Wolfgang von Lengerke, who had fiercely criticized Ganz in recent years for advocating a streamlined, rear-engined car, wrote: "Exceptionally interesting and ideal for the future is the Tatra 77, whose styling expresses roughly how we expect the automobile to look a few years from now."[7] Less than three years earlier, Von Lengerke had dismissed streamlining as a gimmick for racing cars that perhaps, in another hundred years, might find its way into normal passenger cars. The editors of *Motor und Sport*, with their rigidly conventional attitude, had brushed aside the swing axle at first as well. Now that the auto industry was moving over to it en masse, Von Lengerke openly praised Josef Ganz's role as a trailblazer: "Rumpler and above all Dipl.-Ing. Ganz are among its most important pioneers and most energetic advocates."[8]

Ganz's influence was everywhere at the Berlin motor show of March 1934, but in the German newsreels he was not portrayed as its most important figure. At an evening showing in a Berlin cinema, Ganz watched as much was made of the opening speech by Adolf Hitler. The commentator's voice cried out bombastically: "The exhibition was dominated by the will of the Führer: create the German Volkswagen!"[9] Hitler was in a position of power from which he could add strength to Josef Ganz's Volkswagen dream, but the champion of the People's Car was not named in the speech. In *Motor-Kritik*, Ganz analyzed how his demand for a Volkswagen costing only a thousand Reichsmark, now taken up by Hitler, could be met. He wrote that at the IAMA various models had been on view that formed a good basis for the development of the ultimate Volkswagen. Closest to the target retail price was the little Framo Piccolo, on offer for RM 1,225, which "could almost be referred to as a Volkswagen." Next came the Bungartz Butz Touren-Wagen for RM 1,450 and Bungartz's Cabrio-Limousine for RM 1,550, the Standard Superior for RM 1,590, and the Hansa 400 for RM 1,650. All these models were built in basically the same way as the Maikäfer: a lightweight structure with a backbone chassis, rear-mounted engine, and independent suspension with swing axles. Ganz anticipated that in all likelihood an improved version of the Framo Piccolo would be put on the market as a Volkswagen for one thousand Reichsmark, but he wrote: "Beside it, though, we need a slightly larger and more powerful type, which could now almost certainly be developed from, for example, the Hansa 400 or the Standard Superior." That evening, Ganz laid down guidelines for the building of this "luxury Volkswagen" with a retail price of RM 1,300. The car would have a rear-mounted, two-cylinder, two-stroke, 0.6 liter engine, either water cooled or air cooled; streamlined bodywork based on designs by Paul Jaray; and a total weight of 500 kilograms (1,102 pounds). Ganz had in mind a wheelbase of 2,300 millimeters (90½ inches) and a track width of 1,100 millimeters (a fraction over 43 inches) at the front and 1,000 millimeters (nearly 39½ inches) at the rear. The car would need to reach a maximum speed of fifty miles per hour, with fuel consumption of between thirty-three and forty miles to the gallon. "Such a model could today be built to order; a year's development would be needed to make it ripe for serial production."[10]

Talks with Hitler

In his article about the IAMA and the Volkswagen, Josef Ganz offered both government and industry a piece of advice: "We urge the putting together of a neutral design team, as a study group to take on the preparatory work, so that the quest for a Volkswagen can reach its goal more quickly."[11] Ganz was unaware that plans for such a team were rapidly taking shape behind the scenes. Discussing the matter with Adolf Hitler at the motor show, Jakob Werlin had proposed appointing Ferdinand Porsche to lead the project. Werlin had the firm impression that Hitler would have endorsed any engineer he had proposed, although Porsche was definitely a name well known to the Führer. This had nothing to do with Porsche's qualities as an auto engineer but with the fact that during the 1914-18 war he had developed the world's first motorized anti-aircraft gun, the Hunderter. Hitler was so impressed by the weapon—which he had seen in action on the battlefield during his time as a serving soldier—that almost twenty years later he still knew every detail of its design. This impressed Werlin to no end. The Hunderter was based on the C-Zug or "C-Train," its hybrid propulsion using both gasoline and electricity, and it was maneuverable enough to be driven across any kind of terrain and set up anywhere. Porsche and Hitler had in fact met once before, in March 1933, when Porsche, at the request of the directors of Auto Union as part of its efforts to secure a subsidy, had explained to Hitler the design of his racing car. Hitler now responded to Werlin's suggestion by saying: "Bring him with you to Berlin some time."

In March 1934, shortly after Adolf Hitler discussed the Tatra 77 at length with Hans Ledwinka in his room at the Kaiserhof Hotel, Ferdinand Porsche was invited there to speak to the German chancellor. In the lobby, he glanced around and soon spotted his old acquaintance Jakob Werlin, who was a little nervous. Werlin had spent months trying to engage an independent engineer, thwarted at every turn by the auto industry. Ferdinand Porsche was his last hope. Porsche greeted Werlin and they shook hands. "Hitler should be here at any moment," Werlin said, and he began pacing up and down the lobby, Porsche in his wake. "He'll explain to you himself what he wants. But I can tell you now that it's about the Volkswagen; he wants at any cost to get it built." The two men went to sit in a corner of the hotel lobby where Hitler drank tea at the same time every afternoon. It was remarkably busy; all kinds of inquisitive people arrived. The Kaiserhof Hotel's head waiter, it later transpired, was in the habit of secretly "renting out" the best seats close to Hitler's table. Hitler soon came in and went over to sit with Porsche and Werlin. The waiter appeared immediately, politely greeted his goldmine Herr Hitler and poured tea. Hitler got down to business straight away, telling Porsche about his plans. The Volkswagen must have four seats, to accommodate a family, and an air-cooled engine so that it could be left outside in winter without the risk of being immobilized by a frozen radiator. The possibility of developing a version of the Volkswagen for use as a military vehicle must be looked into as well. "I fought in the war myself," Hitler digressed briefly, "and I know from my own experience how difficult it is under those circumstances to get hold of antifreeze." Hitler laid his rough sketch of a few months earlier on the table. He had

Sketches made by Adolf Hitler in the Kaiserhof Hotel, Berlin, 1934

brought a sketchbook with him, and within a few minutes, still talking, he made six sketches of how he imagined the Volkswagen should look.

Hitler's drawings do not attest to a great deal of imagination. They look very much like the Mercedes-Benz 130H that he had seen a short time before at the IAMA. At first he drew air intakes for the engine on the sides near the front, then crossed them out and moved them to the back. Finally, he sketched a side view of his adapted Mercedes-Benz 130H, with a short front and an elongated rear, and outlined with his pencil on top of the drawing the contours of a water droplet, bifurcated at road level, and next to it a view of the same streamlined shape as seen from above. Once again, the design was in line with suggestions by Josef Ganz, who had written in the most recent issue of *Motor-Kritik*: "The type 130, by contrast, has been given a rear-mounted engine. We can hope that the appropriate styling for the car, the truly streamlined superstructure, is yet to come."[12] Like Hitler, Daimler-Benz listened to Ganz's criticism of the body shape and the "outboard motor" layout of the 130H when it designed the more streamlined coupe 150H with a mid-mounted engine for the 2,000 km through Germany rally, to be held that July.

Hitler looked up from his sketchbook and announced his final stipulation: the Volkswagen must not cost more than a thousand

Sketch made by Adolf Hitler in the Kaiserhof Hotel, Berlin, 1934

Reichsmark—the retail price Ganz had been advocating for years in *Motor-Kritik*. Hitler added that mass production in a modern factory, following the example set years ago by his personal role model Henry Ford, must surely make such a price feasible. After his last sip of tea, he got up to leave. He handed his sketches to Werlin, said good day to him and to Porsche, and left for his study. Porsche and Werlin had hardly spoken during the conversation, or rather monologue, which had lasted less than fifteen minutes. Porsche looked rather disheartened. He bowed his head and said to Werlin: "Now I know what he wants."[13] What Hitler wanted was not simple. The new chancellor was absolutely determined that Josef Ganz's Volkswagen idea—for years the subject of fierce debate within the car industry—would become a reality. The guidelines had already been formulated in *Motor-Kritik*. After Porsche and Werlin parted, Werlin sat at the table for a while and wrote at the bottom of the last sheet of paper: "These sketches were made by Hitler on the occasion of a discussion with Porsche and me in the Kaiserhof Hotel (1934). The work of just a few minutes!"[14] In

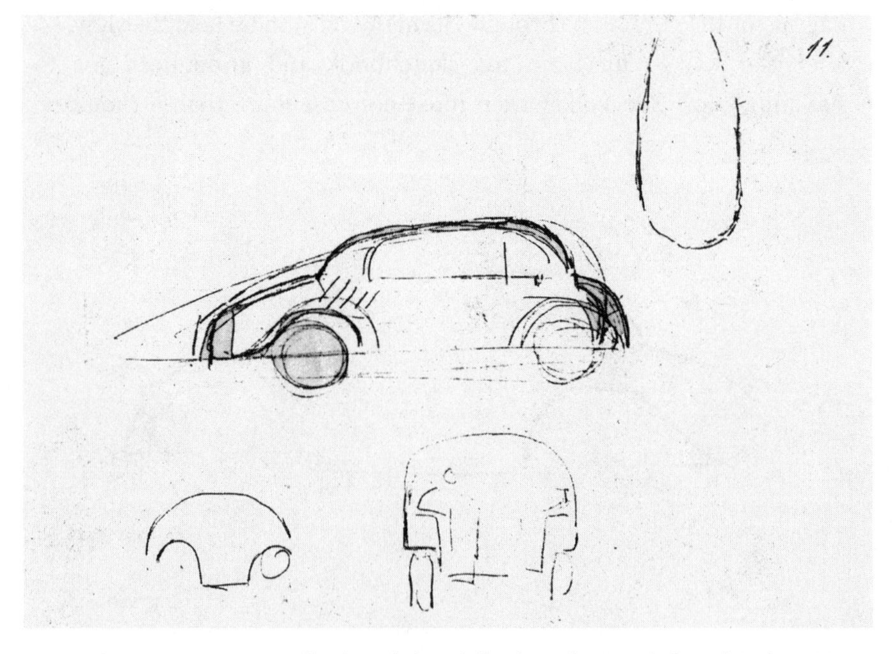

Sketch made by Adolf Hitler in the Kaiserhof Hotel, Berlin, 1934

Sketch made by Adolf Hitler in the Kaiserhof Hotel
with notes by Jakob Werlin, Berlin, 1934

those few minutes, Hitler had laid the basis for the Volkswagen. His plans bore the stamp of Josef Ganz and his magazine, but of course a National Socialist state project could not possibly admit to Jewish sources. On Werlin's shoulders lay the task of making short work of the "Jewish vermin" that were still infesting the upcoming Nazi Volkswagen project.

Ban on *Motor-Kritik*

One afternoon in late March 1934, the large black Bakelite telephone on the desk of the publisher Wilhelm Breidenstein rang. On the line was Adolf Heinz Beckerle, superintendent of the Frankfurt police, who brusquely informed him in no uncertain terms that the publication of the magazine *Motor-Kritik*, which was in his hands, was banned with immediate effect. The message was a bolt from the blue. *Motor-Kritik* had certainly had its conflicts with the industry in the past, but there seemed no reason at all for a total ban by the government. Wilhelm Breidenstein immediately discussed this bizarre turn of events with his brother and co-publisher Hartwig Breidenstein and with the editor-in-chief of *Motor-Kritik*, Georg Ising. Late that afternoon, the three men visited Beckerle for clarification, but the superintendent could add little to his earlier order, except to say that the ban on publication of the magazine had originated in "the very highest circles." The only person who could tell them more was Dr. Jacob Otto Dietrich, the Nazi Party's chief press officer and SS-Gruppenführer in Berlin. That same night, Georg Ising and the Breidenstein brothers set out by car for the capital, a rough and exhausting journey of many hours on well over three hundred miles of unlit provincial roads that passed right through towns and villages. To avoid wasting any time, they took turns driving and kept going practically without a break. Josef Ganz wisely chose not to go with them. His presence at the News Service offices in Berlin might only have added fuel to the fire. He stayed behind in Frankfurt with his girlfriend Madeleine Paqué, tensely waiting for the phone to ring.

The sun was coming up by the time Georg Ising and the Breidenstein brothers reached Berlin. After a quick breakfast and several cups of strong black coffee, the three men reported to the German News Service. Dietrich turned out not to be there that day, but they were let in to see Professor Doctor Alfred Herrmann, member of the board and boss of the head office of the German National Press Association. Herrmann handed them the file relating to the ban on *Motor-Kritik*. Georg Ising opened the thin cardboard folder. It contained only a few sheets of paper, among them a clipping of the opening article "M-K Sorrows of Our Time," from *Motor-Kritik* number 5, of early March 1934. The article had been clipped in such a way that it was impossible to see from which issue it had come. An accompanying text stated that in his article Josef Ganz had expressed disloyal criticism of the opening speech by Adolf Hitler at the recent motor show in Berlin. Those lines turned out to have been written by Jakob Werlin, the man Hitler had put in charge of the Volkswagen project. Georg Ising pointed out to Herrmann that

Above: The new model of the Standard Superior (right) stands next to an Adler, 1933.
Top: The new model of the Standard Superior was featured in the cigarette-card series "Das Auto von Heute" issued by Garbaty Cigarettes, 1936.

Above: The new Mercedes-Benz 130H Cabrio-Limousine outside of the home and office of Josef Ganz, Frankfurt, 1933. Ganz borrowed the car for a test drive.

Top left: Motor-Kritik introducing the new rear-engined Mercedes-Benz 130H, 1933.

Top right: Frits Ostwald, the son of *Motor-Kritik* editor Walter Ostwald, next to the new Mercedes-Benz 130H, in front of the *Motor-Kritik* editorial office.

Above: The backbone chassis of the Mercedes-Benz 130H with what Josef Ganz called an "outboard motor."

Top and middle left: The engine cover of the Mercedes-Benz 130H opened backwards, exposing the water-cooled, four-cylinder engine.

Top right: The front of the Mercedes-Benz 130H housed the spare wheel.

Top and bottom: Motor-Kritik editors Georg Ising and Walter Ostwald join Josef Ganz on a test drive with the Mercedes-Benz 130H through the Black Forest, 1933.

Top: One of the NSU Type 32 prototypes developed by Porsche, 1933. This model featured a body made of wood covered in artificial leather with metal fenders, 1933.
Middle: This Tatra V570 prototype with its streamlined body and rear-mounted, air-cooled, two-cylinder engine was intended as a "Volkswagen," 1933.
Bottom: The prototype of the streamlined Tatra 77 with a rear-mounted, air-cooled, V8 engine, 1933. In 1934, Tatra started the production of this luxurious 77 model.

The stand of the Standard Fahrzeugfabrik at the Berlin motor show, displaying a streamlined sports version of the Standard Superior, 1934.

This car displayed by Standard was one of several sports cars built on the chassis of the Standard Superior, Berlin, 1934.

XIV. JAHRGANG NR. 5 ANFANG MÄRZ 1934

MOTOR-KRITIK

VERLAG FRANKFURT AM MAIN, BLÜCHERSTRASSE 20/22. EINZELHEFT RM 0.60

Framo „Piccolo"

Ein neuer deutscher Kleinst-Wagen mit Schwingachse und Heckmotor

Ausstellungs-Vorschau

Top: Motor-Kritik introducing the Framo Piccolo, which was one of several small cars that resembled the Standard Superior, Berlin, 1934.

Middle: The Framo stand at the motor show, featuring the Framo Piccolo, Berlin, 1934.

Bottom: The rear-engined Hansa 400 also closely resembled the Standard Superior.

The new model of the Standard Superior "streamlined limousine" was again offered for the original price of 1,590 Reichsmark, Berlin, 1934. The brochure for the car proclaimed it to be "the fastest and cheapest German Volkswagen."

The stand of the Bungartz company featured three versions of the new Bungartz Butz, small cars developed according to the patents of Josef Ganz, Berlin, 1934.

HANSA-LLOYD
60
RM. 1450
BUNGARTZ & CO.
BUTZ
MÜNCHEN

Two days after its official introduction in Prague, the streamlined Tatra 77 was also presented at the Berlin motor show, 1934. Next to the car, Tatra put an impressive 77's rear-mounted V8 engine block arrangement on display.

Top: A *Motor-Kritik* edition featuring the Berlin motor show with a streamlined car design by Belá Barényí pictured on its cover, 1934.
Bottom: The air-cooled V8 engine of the Tatra *77*, Berlin, 1934.

Top: A new model by DKW showing more streamlined bodywork, Berlin, 1934.
Middle: Sports car by BMW with the headlights built into the grille for better streamlining, Berlin, 1934.
Bottom: A front-wheel-drive DKW at the motor show in Berlin, 1934.

A Hanomag chassis with swinging axles and a low center of gravity, Berlin, 1934.

Top: The streamlined Mercedes-Benz 500 Kompressor with, in the background on the left, two busts of the auto pioneers Gottlieb Daimler and Carl Benz, Berlin, 1934.
Bottom: A woman sells photos of the cars on display at the Berlin motor show, 1934.

Top: The streamlined coupe Mercedes-Benz 150H competing in the 2,000 km rally through Germany, 1934. The design shows a striking resemblance to the later VW Beetle.
Middle left: On the advice of Josef Ganz the "outboard motor" layout of the 130H was changed to a mid-mounted engine on the 150H, 1934.
Middle right plus bottom left and right: Computer reconstruction of a streamlined Volkswagen design by Josef Ganz, in part based on the work of streamlining pioneer Paul Jaray. The bodywork sits on top of a backbone chassis similar to the one used on the Standard Superior, 1934.

the article—the sole evidence on which the ban on *Motor-Kritik* was based—had been published five days before Hitler delivered his speech at the IAMA. Ising had brought recent issues of the magazine with him and he managed to convince Herrmann he was right. Josef Ganz could not possibly have been criticizing what Hitler had said. Herrmann accepted his version of events and immediately withdrew the publication ban, but he forbad Josef Ganz from writing for the magazine and the editors from publishing anything further about him.[15] This ban applied to the entire German press. Furthermore, Ganz was forbidden to link his name to any technical innovations. The individual called Josef Ganz was silently removed from the public domain in March 1934. The falsification of history was complete.

Later that morning, Georg Ising rang Josef Ganz in Frankfurt with both good news and bad. It was the latest of many blows for Ganz. That very morning, he had received a letter from VDO Tachometer, a manufacturer of dashboard instruments for which he had been working as a consultant since October, saying that "to their regret" the directors were forced to end their contract because of "the need to save money."[16] In less than a year, Ganz had lost everything: his contracts with BMW, Daimler-Benz, VDO Tachometer, and *Motor-Kritik*. Jakob Werlin had now delivered the death blow by breaking the public connection between Ganz's name and the Volkswagen.

Dolly Prevents an Attack

On April 11, 1934, several weeks after the conversation between Hitler, Porsche, and Werlin in the Kaiserhof Hotel, an official gathering took place to discuss "the creation of the Volkswagen." Those present included representatives of the Chancellery, the Ministry of Propaganda, the Ministry of Economic Affairs, and the RDA. The chairman reminded them of Hitler's speech at the IAMA in Berlin, in which he had called for the building of a German Volkswagen. Points of departure were the maximum price Hitler had proposed of one thousand Reichsmark and running costs of six pfennigs per kilometer, or 3.73 pfennigs per mile (100 pfennigs = 1 Reichsmark). The car would have to be capable of a maximum speed of fifty miles per hour with fuel consumption of between forty-seven and fifty-nine miles to the gallon, have room for three adults and one child, and be mass produced. During the meeting, the possibility arose of making it a three-wheeled car, with two front wheels and one rear wheel, featuring "a natural streamlined shape." The RDA representative said once again that the industry was already at work on a Volkswagen and requested that its development be left to existing manufacturers, to which the representative of the Chancellery replied that the industry had only ever built cars that were far too expensive.[17]

The question of whether the Volkswagen should have three wheels or four was put to Hitler at the beginning of May. He expressed a preference for "the four-wheeled Volkswagen."[18] He had become attached to the design Josef Ganz had described in *Motor-Kritik*. The nickname Maikäfer for the four-wheeled Volkswagen with a rear-mounted engine had penetrated so deeply into the German psyche that Hitler, speaking to an engineer about the shape of the car at a meeting in the Ministry of Propaganda, announced: "It should look like a Maikäfer. You only need to look at nature to know how streamlining ought to be applied." The engineer answered that the May Bug's hard shell was intended not for flying but only "to protect its wings" as a kind of "garage." It mostly scuttled about on the earth. Hitler stood up and interrupted the engineer, saying sarcastically: "What does the Maikäfer do with its garage when it flies, then? Does it leave it standing on the ground?"[19]

In early April 1934, as discussions about the Volkswagen were getting into their stride, Ganz filed his writ against Paul Ehrhardt, Gustav Röhr, and Bernhard Stoewer at the court in Frankfurt, a document of more than seventy pages. Several days later, when he arrived home in his Maikäfer late one evening and turned onto the drive next to the villa on the Zeppelinallee, his German shepherd Dolly, barking loudly, leapt out of the car and attacked a man who was hiding in the dark behind the garden railings. The intruder hit Dolly hard on the head with a cosh he was holding ready and fled. In the bedrooms, the lights came on. Madeleine

Paqué and her mother had been startled awake by the barking and the commotion. They ran out and found Josef kneeling next to Dolly, who was lying unconscious on the ground. He picked up the dog and took her inside, then went back out to investigate further. In the garden and on the balcony he found several cigarette butts. He concluded the man must have been lurking there for some time. He picked up the stubs and took them inside. The identity of the intruder remained a mystery. No one could sleep that night. The only sound was the tapping of a typewriter in the study, where Josef Ganz immediately wrote a full account of events for the police. Early next morning, he got into his Tatra and drove to police headquarters in Frankfurt. There he reported the attack and handed the officer on duty his statement and the cigarette butts he had found on the balcony. The policeman took the stubs, flung them to the floor, and trod on them.[20] This Jew's story was of no concern to him.

Ehrhardt Delays Court Proceedings

Paul Ehrhardt, who had stepped forward as spokesman in response to Ganz's charges, did his best to obstruct progress of the case in the weeks that followed. On April 23, 1934, he wrote to the court that the complaint submitted by Josef Ganz was "so extensive" that a "comprehensive response" was impossible within the fourteen days available. In his letter, he pointed to three "fundamental conclusions" about which he would provide further information in his next letter. Under point one, Ehrhardt wrote that he had "in no way publicly libeled" the plaintiff but had merely reported, in a concise form, the facts about Josef Ganz as he knew them and that he had fully documented all the allegations made. "It is now the task of the private plaintiff to call into question the evidential value of these documents." Under point two, Ehrhardt wrote that he had handed "this written document in multiple copies exclusively to the district leadership of the Nazi Party in Frankfurt," so as to "give the appropriate civil servants the opportunity to check the facts and understand their implications." Under point three, he wrote:

> I can assure you that the occasion for taking these steps is not the campaign of destruction and libel the private plaintiff has been waging against me for three years, since I take the view that the private plaintiff will bring about his own downfall with such fighting methods and such a character (quite apart from of the fact that the private plaintiff is a Jew). I'm far more concerned to pillory journalistic vermin like the private plaintiff as quickly as possible as a matter of public interest, in such a way that the profession can take measures to render it harmless. If such a course of action is a punishable offence, then I request my punishment.[21]

Two days after submitting this first brief statement, Ehrhardt asked the court to extend the period available to him by at least four weeks, given that over the past couple of months he had been "extraordinarily overburdened with work" and was about to leave on "a long business trip," which made it impossible for him to answer a seventy-page legal document that required "thorough study." Ehrhardt signed his letter with a firm "Heil Hitler!"[22] Shortly after putting in his request, he left with his wife Elfriede in a small private plane on a business trip to Switzerland. Four days later, a drama took place there that would further delay court proceedings. Taking off from Basel airport on April 29, 1934, Ehrhardt's plane crashed, killing his wife. He survived the accident but was taken to Basel hospital with a number of complex fractures. In late May, when the court in Frankfurt was expecting finally to receive a written response from Ehrhardt in the case, Gustav Röhr delivered a statement saying that Ehrhardt had been in a plane crash and was "in traction with complicated bone fractures." Röhr argued that "further delay can therefore hardly be avoided," since "neither of the other defendants is sufficiently, or indeed in some respects at all, familiar with the material."[23] Röhr wrote that Ehrhardt would not be fit to make the return journey from Switzerland until July—although in fact he had been flown back to Germany shortly after the accident, on the personal orders of Hermann Göring.[24]

In contrast to the long-drawn-out court case against Ehrhardt, Röhr, and Stoewer, the case Tatra had brought against Ganz in December 1933 for patent violations was moving forward. Its claims had been rejected by the Patent Office in Berlin in late March 1934.[25] Tatra had appealed a fortnight after the verdict, but its appeal was rejected in May 1934.[26] For the time being, the dispute with Tatra seemed to have been settled, but the company was preparing to submit a second complaint. This time Tatra would claim that Ganz's patent DRP 587409[27] violated its patent DRP 469644.[28] Madeleine Paqué suggested to a nerve-wracked Josef that they should take advantage of the victory over Tatra and the delay to the case against Ehrhardt, Röhr, and Stoewer by going away on holiday for a few weeks. Josef agreed. After all the problems he had faced over the past year, he was keen to drive to Switzerland for a rest in the mountains at his Uncle Alfred's villa. He could also take the opportunity to visit Paul Jaray, who that same month had finalized a bodywork design for a four-seater streamlined Volkswagen, to be built on the chassis of the Bungartz Butz.[29]

Resistance to Hitler

In late May 1934, another official meeting took place to discuss the German Volkswagen, this time at the Manufacturers' Commission, part of the RDA. Firms including Adler, Auto Union, BMW, Daimler-Benz, and Opel were represented. During the meeting, it was decided, reluctantly and with considerable reservations, to commission Ferdinand Porsche to develop the car.[30] In the weeks that followed this meeting, more and more letters began to circulate between the directors of German manufacturing firms and the RDA, with car builders rejecting Hitler's ideas. Director Popp of BMW regarded the incomes of 90 percent of the German people as far too low to permit them to buy a car. They were destined to travel by "bicycle, streetcar, and train." For the general public to benefit from "the blessings of motorization" there was only one suitable vehicle: "the omnibus!"[31] If any manufacturer was already in a position to mass produce automobiles, then it was Opel, capable of producing five hundred a day. A double-shift system could increase that to a thousand.

Popp believed it would cost at least a hundred million Reichsmark to build a new factory comparable to Opel's, and that, he said, would be "economically absurd," since the German market was not large enough to warrant it. He wrote to Wilhelm Kissel, director of Daimler-Benz: "I regard it as completely wrong for the RDA to take on this problem by giving Dr. Porsche or anyone else the task of designing such a car, since that means the German auto industry will be admitting it has been asleep up to now and has not been building the right kind of small cars."[32]

Opel joined in, making frantic efforts to convince both the RDA and Hitler personally that it had the capacity to build the cheapest fully fledged small car. James D. Mooney, the head of overseas operations for Opel's mother company, General Motors, visited Hitler to argue the case.[33] Efforts to resist the coming of the Volkswagen were fruitless, however. Hitler pushed his plans through, and in early June 1934 the RDA presented Porsche with a proposal for a contract that would have to be signed before the month was out.[34] Unaware of government plans, Josef Ganz and Madeleine Paqué packed a few clothes into two suitcases, and on June 11, 1934, they set off for the Swiss border in their Tatra for a holiday from which they would never return.

CHAPTER FOURTEEN

NIGHT OF THE LONG KNIVES

Josef Ganz and Madeleine Paqué got up early on the sundrenched morning of July 1, 1934. Madeleine lovingly wished Josef a happy thirty-sixth birthday, and they enjoyed a delightful breakfast with a view of the Swiss Alps. They had spent almost three untroubled weeks driving their Tatra through the mountains, far from all the feuding in Germany. They had swum in icy mountain lakes, visited the impressive Morteratsch glacier and other sightseeing highlights of the Bernina Range, and driven their Tatra up the Piz Bernina, the highest mountain in eastern Switzerland. At the parking lot near the Morteratsch glacier, Josef Ganz met a German couple that had undertaken the voyage in an old Hanomag 2/10 PS Kommissbrot, just as he had done seven years earlier, and he told them enthusiastically about his own experiences with the car. After touring the Alps, Josef and Madeleine visited his uncle and aunt, Alfred and Valerie Ganz, in St. Niklausen, and from there they drove on to see Paul Jaray in nearby Luzern. Ganz and Jaray discussed the latter's design for a four-seater streamlined body to be fitted to the chassis of the Bungartz Butz, which he had completed in May.[1] It would be a teardrop-shaped two-door car, with fenders and headlamps fully integrated into the bodywork. Bungartz wanted to combine Josef Ganz's design for the chassis with Paul Jaray's streamlining to create an "uncompromising Volkswagen"—precisely the kind of car Hitler had called for in his speech at the IAMA. Ganz had recently test driven a streamlined prototype that Jaray had created for Daimler-Benz not long before and closely resembled this design, with one important difference in that it was a front-engined car. Jaray took Josef Ganz and Madeleine Paqué on a tour of the countryside around Luzern in his prototype, the highlight of the ride being the magnificent panorama from the top of Mount Pilatus, at 6,995 feet above sea level. Now, sadly, their trip was coming to an end. They wanted to try to reach Frankfurt by evening to celebrate Josef's birthday in style with family and friends. Immediately after breakfast, Ganz paid the hotel bill, loaded the suitcases onto the back seat of the Tatra, and set off with Madeleine for the German border.

Massacre in Germany

They drove right across Switzerland on beautiful country roads from village to village, but they were somewhat surprised at the commotion in the streets of the first few places they came to, so they bought one of the extra editions of the newspapers that were on sale everywhere. Still planning to be in Frankfurt by evening, Josef drove on immediately while Madeleine read aloud from the paper. The front page was full of bizarre and shocking reports of what had just happened in Germany. Apparently on the night of June 29 and into the next day, a massacre had taken place. In a campaign organized by the Nazi government, SS death squads had killed dozens of members of the SA, including its leader, Ernst Röhm. The Gestapo, presumably on the orders of Hitler, Göring, and Himmler, had eliminated several other opponents of the regime as well, along with a number of the leadership's own personal enemies. Alarmed by the reports, Josef and Madeleine decided not to drive straight on into Germany but to find a hotel and try to contact Georg Ising in Frankfurt the following day.

On the morning of July 2, Ganz rang the editorial office in his house on the Zeppelinallee in Frankfurt from a payphone. Georg Ising answered, and Ganz asked him what in God's name was going on in Germany. Ising replied rather curtly, saying that at home they had not spoken of the events, but it was good that Josef had called. Things were extremely quiet at the office, and he would be well advised to stay in Switzerland for a while "to recover from his lung infection." Josef Ganz was surprised—he had not been suffering from a lung infection—but he quickly understood that Ising felt he could not speak freely and was giving him a coded warning. Ganz thanked him, said goodbye, and hung the heavy Bakelite receiver back on the metal hook. From the open door of the phone booth, he saw Madeleine standing waiting next to the Tatra with two suitcases of clothes on the back seat. Everything else they owned was in Germany. In Frankfurt, Georg Ising hung up, hoping Josef had understood.

Ising had been present in person when, on the night of June 29, the Gestapo had come to the door to arrest Josef Ganz. But he could not speak openly about that. A black car, a Horch, had been parked for several days on a stretch of wasteland across the street. At the wheel was a chain-smoking Gestapo man, watching the house. He had set up a makeshift phone tap so he could listen to any calls made or received by the *Motor-Kritik* editorial office. From time to time he got out of the car and stopped people coming and going from the house to ask them where Josef Ganz was and when he would be back. They all kept their mouths firmly shut.[2]

Josef Ganz decided to stay in Switzerland while Madeleine Paqué took the train to Frankfurt to be with her mother, whose health was precarious. But the situation in Germany did not improve. The massacre of June 29, 1934, which Hitler later named the Night of the Long Knives, marked a turning point in Nazi Germany. On August 2, President Paul von Hindenburg died. He had been suffering from dementia for some time and his senility had allowed Adolf Hitler to do pretty much whatever he liked as chancellor. He now took advantage of the president's death by immediately, the same day, merging the offices of president and chancellor. Von Hindenburg's body was barely cold before Hitler managed to get himself thrust into the position of Germany's Führer, its one and only leader.

His most important political opponents had been removed during the Night of the Long Knives, and the army now swore total obedience to him. Germany had officially become a National Socialist dictatorship. In Switzerland, Josef Ganz was fast running out of money as a result of his enforced stay in hotels, so he tried as best he could to start working again.

Meeting with Porsche

Ganz had been banned from writing for *Motor-Kritik* on the orders of the Nazi Party, but Georg Ising nevertheless published several articles by him that summer, without any byline. One of those contributions, which appeared in the magazine in mid-August, was a report on the legendary Klausen Rennen of August 5, 1934, in which racing cars and motorcycles from Switzerland and beyond competed over a 13.4-mile course through the mountains.[3] The route ran past the Swiss mountain village of Linthal and through the Klausenpas, along treacherous gravel tracks with 136 bends and changes in altitude totaling 4,058 feet. The racing cars included two German teams sponsored by Hitler, one representing Auto Union and the other Mercedes-Benz. At the wheel of the Auto Union Type A was the legendary racing driver Hans Stuck, who had won the Grand Prix on Germany's Nürburgring in the same car on July 15. The equally famous Rudolf Caracciola raced in the Mercedes-Benz W25. Ganz met several old acquaintances during the Klausen Rennen. Ferdinand Porsche and the racing team's leader Baron Klaus-Detlev von Oertzen were there with the Auto Union team, and support for Mercedes-Benz came from the team leaders Alfred Neubauer and Hans Nibel. It was an extraordinarily competitive and thrilling race. Josef thoroughly enjoyed himself, and he put together a lengthy photo-reportage that was published in *Motor-Kritik* alongside his account of the race. Caracciola achieved the fastest time in his Mercedes-Benz and set a new class record by covering the route in 15 minutes, 22.02 seconds, with Stuck hot on his heels in the Auto Union car, coming in at 3.02 seconds slower.

After the race, Josef congratulated Porsche on the convincing performance of the Auto Union racing car and talked with him about its design, as well as about the Volkswagen and the situation in Germany. Ganz still greatly respected Porsche's work as an engineer, and Porsche had a similar respect for Ganz and his progressive ideas as a pioneer of the Volkswagen.[4] Porsche's design bureau had signed a contract with the RDA on June 22 to develop three prototypes of the German Volkswagen. At Porsche's office in Stuttgart, his staff was busily working on the first designs. They had only ten months in which to produce the prototypes. Ferdinand Porsche's own time was taken up with the races in which Auto Union cars were participating, and that summer he traveled all over Europe with the racing team to Grand Prix circuits including Monaco, Montlhéry in France, and Monza in Italy. From Switzerland, the Auto Union and Mercedes-Benz teams moved on to the Coppa Acerbo Grand Prix, due to take place on August 15 in Pescara in Italy. There each of the two manufacturers raced two cars, with William Sebastian joining Stuck as a driver for Auto Union and the Italian driver Luigi Fagioli joining Caracciola to drive for Mercedes-Benz. During the race, a ferocious battle developed between Caracciola, Stuck, Fagioli, and the Italian driver Achille Varzi in an Alfa Romeo. Fagioli won in his Mercedes-Benz and was personally congratulated by Jakob Werlin, who had come to Italy to watch the Grand Prix.

Secret Trip to Germany

Unable to return to Germany, Ganz stayed on in Switzerland. Later that month, he visited the Internationale Alpenfahrt, an Alpine race that again pitted dozens of racing cars from different countries against one another, including models from Adler, Bugatti, Citroën, Ford, and MG. Here too Josef saw many friends and acquaintances from the auto industry, including Georg Ising, who had traveled from Frankfurt for the rally. Among the wide range of cars taking part was a unique streamlined model on a Ford chassis with a V8 engine. Ganz wrote a report of the rally, and again it was published in *Motor-Kritik* without a byline, along with his photo-reportage.[5] Ising had updated Ganz on developments in Germany and brought important mail to Switzerland for him, including a letter from the lawyer Dr. Jur. Theodor Hartherz in Frankfurt, who informed Ganz that on August 30 the court had delivered its verdict on his case against Paul Ehrhardt, Gustav Röhr, and Bernhard Stoewer. Ganz was unsurprised to hear that all three had been found guilty, at least in theory. Ehrhardt and Röhr would have had to serve terms of up to six months in jail and Stoewer pay a fine of a thousand Reichsmark—were it not that, astonishingly, all three had been exempted under the new Law on the Awarding of Immunity from Prosecution, introduced five days after Von Hindenburg's death.[6] The law granted amnesty to defendants who had been sentenced, prior to its introduction, to fines of up to a thousand Reichsmark or prison terms of up to six months but had not yet paid or been jailed.[7] By delaying their case for several months, the three men had managed to escape punishment. Ganz was flabbergasted and asked his lawyer to mount an appeal, but the verdict was a clear sign that under present circumstances it was impossible for him to return to Germany. The country he had defended with his life during the First World War had become a dictatorial rogue state governed by ruthless opportunists.

Josef Ganz had no choice but to try to build a new life in a foreign country, but he wanted at all costs to try to save his priceless documents, which were at his house in Frankfurt, from the clutches of the Nazis. It would mean mounting an extremely dangerous rescue effort, inside the lion's den. Ganz kept close track of developments in Germany and waited for his moment. He eventually decided that his secret trip to Germany should be timed to coincide with the annual Nuremberg Rally, due to be held that year on September 5-10. It would be a mass event, with tens of thousands of Nazi Party members of all ranks arriving from every region of Germany in overcrowded trains, buses, and army trucks. Ganz decided to gamble that controls at the borders and in the rest of Germany would be more lax than usual during that period. Josef Midinet, an old family friend in Frankfurt, offered to help. It would be too dangerous for Ganz to make the journey in his own car, so in early September Midinet drove a Röhr Junior Limousine from Frankfurt to Switzerland and met him there. Together they drove to the German border on September 5, just as tens of thousands of Nazi Party members were gathering in Nuremberg. The young border guard simply raised the barrier and waved most people through without any kind of check, Ganz and Midinet among them. They were in Germany.

The journey was uneventful until, in the middle of Germany, the engine began to falter and then fell silent. They could not get the car going again. Ganz opened the hood and saw that condensation had formed in the distributor cap in the wet weather and caused a short circuit. He tried the electric starter motor a few more times, then cranked the engine over and over, in vain. Ganz and Midinet were considering their options when they heard the sound of loud male singing. It was getting closer. Ganz looked up in alarm and saw through the small, slightly misty rear windshield a platoon of SA soldiers marching directly toward the car. Before Ganz had time to think, the officer in charge, seeing the hood up, asked if they had broken down. Josef decided to do his best to play along and said that yes, the engine was refusing to start because of condensation in the distributor cap. The soldiers offered him a push-start. Ganz immediately closed the hood and got in. Smiling cheerfully, the soldiers slung their weapons over their shoulders and soon, with twenty strong pairs of hands behind it, the car started to move. In a cold sweat, hoping against hope, Ganz released the clutch. The car jolted and the engine came to life, sputtering and bubbling. He tramped the gas pedal a couple of times, shouted a few heartfelt words of thanks through the side window, put the car in gear and drove off at speed. Watching the soldiers waving happily as they receded from sight in the rear-view mirror, a feeling of anxiety and melancholy overcame him as he contemplated how these young men, so friendly by nature, had been incited to race hatred and a thirst for war simply because they despaired of ever finding work. All their hopes for a better future were invested in the Nazi regime.

They reached Frankfurt a few hours later. Ganz waited in the car, engine running, while Midinet went into the villa on the Zeppelinallee. If the Gestapo were waiting for him he could attempt a getaway. Midinet waved to indicate the coast was clear, so he put the Röhr Junior in gear and drove through the metal gates onto the drive. Jumping out of the car, he walked briskly inside to an enthusiastic welcome from Madeleine. They had not seen each other for weeks. Georg Ising was extremely happy to see Josef again and congratulated him on the success of his courageous mission. Ising and Midinet helped Ganz to load his most important documents and technical drawings into the car. Georg Ising's wife Gustel had quickly made lunch and brewed tea, which they enjoyed together in the back garden. Ganz would have loved to remain right there, in this stolen moment, with his beloved Madeleine and faithful friends, in the half-overgrown garden in beautiful sunny late-summer weather. Life could be so good. But the danger was such that he had to leave Nazi Germany as soon as possible. He thanked Georg Ising for his help and friendship and said goodbye to Madeleine, not knowing where, when, or in what circumstances they would see each other again. He stroked Dolly's head. The faithful German shepherd that had saved his life a few months before would die less than three weeks later. With a lump in his throat, Josef Ganz got into the car, its back seat buried under piles of bundled documents. As a final item on their agenda, Ganz and Midinet drove to the bank. Josef wanted to withdraw the savings he so badly needed, but his account turned out to have been frozen. Fearing their visit to the bank might have attracted attention, Ganz and Midinet drove nonstop for hours until, in pitch darkness, they reached the depths of the Black Forest. There, in a remote village, they found a tiny hotel where they could spend the night. Midinet went in first, to chat with the owner and check if it was a safe place to stay. When he was certain no Nazis were about, he fetched Josef from his hiding place behind a corner building on the other side of the street.

Ganz and Midinet left before sunrise the following morning, to the surprise of the friendly hotel owner. They covered the final few dozen miles by driving across open country to avoid the larger German border posts. "Long live the swing axle," cried Ganz, in high spirits, but he was anxiously watching the fuel gauge as the car slowly sputtered along. The Swiss border checkpoint came into sight just as the car began to jolt and falter, and after a few violent lurches and a loud bang the engine stopped, and the car rolled down the hill to come to rest right at the border. The guard at the little border post thought this extremely odd, and indeed suspicious. To convince him of the importance of their mission, Ganz showed him the documents and a wooden scale model of his Volkswagen. He also supplied the name of a good friend, an influential man who worked at the Georg Fischer AG steel factory in the Swiss town of Schaffhausen. The guard went indoors and rang the friend, while Ganz and Midinet waited on tenterhooks. A few minutes later he came out and gave them permission to drive on. Smiling broadly, they pushed the car past the checkpoint to a nearby gas station, where the trusty Röhr Junior was filled and they stopped to drink a strong cup of coffee to calm their nerves, glad to have succeeded in their mission.[8]

Midinet dropped Ganz next to his own car and the two men parted. Midinet drove back to Frankfurt, while Ganz, when his Swiss tourist visa ran out, moved on to France. From Paris, he continued to stay in close touch with Frankfurt. Georg Ising tried to support him as best he could, keeping him up to date with the latest developments and regularly sending him German automobile magazines including *Motor-Kritik*, *Motor und Sport*, and *Allgemeine Automobilzeitung*. People who worked with Ganz were appalled by his forced departure. In late September, Ising had spoken to Walter Ostwald, who told him: "Josef Ganz deserves historic recognition for lifting the German auto industry out of stagnation."[9] Georg Ising did everything he could to keep *Motor-Kritik* going purely as a specialist magazine, avoiding any involvement in politics, but because of Nazi censorship he could not publish another word about his good friend, the former editor-in-chief of the magazine and originator of the Volkswagen. All across the German media, the influential engineer's name had slipped silently into oblivion, but he continued to have a powerful influence on the car industry.

Meanwhile, the German media were reporting that Ferdinand Porsche had been commissioned by the RDA to develop a German Volkswagen and was already hard at work on the project in Stuttgart.[9] In October 1934, thinking about the propaganda that would be used to sell the car in the future, the RDA decided to secure rights to the name Volkswagen, which was now popping up with great regularity in auto manufacturers' advertisements. It registered the name at the Patent Office in Berlin,[11] forcing all car companies to stop using it. The Standard Fahrzeugfabrik had to print a new brochure. The cover that had shown a German family standing next to "the fastest and cheapest German Volkswagen" was replaced by a photograph of the Standard Superior bearing the caption: "The most reliable, comfortable, and economical small car."[12]

Murder Plots

That month, Josef Ganz paid his annual visit to the Paris motor show, the Salon de l'Automobile, where he spoke with old friends and famous figures from the German auto industry. On the morning of October 10,

FRANKFURT
Heftrand

1 2 3 4 5 6 7 8 9 10 11 12 13 14 15 16 17 18 19 20

Name: (bei Frauen auch Geburtsname) Ganz,
Wohnung: Frankfurt/Main, Zeppelin-Allee 85,
Bildvermerk
Vorname: Joseph,
Geboren: 1.7.98 zu Budapest,
Familienstand:
Beruf: Dipl.Jng.,
Deck- Name: Adresse:
Glaubensbekenntnis:
Personalakten:
Staatsangehörigkeit:
Politische Einstellung:

Zeit	Ort	Tat	rechtskräftige Gerichtsentscheidung	Staatsanwaltschaft und Aktenzeichen	Polizeiakten

Strafen:

Front of Frankfurt police registry card for Josef Ganz, 1934 he had a meeting at his room in the Hotel Osbourne with a friend, Herr Goldstein, the former chairman of Germany's National Association for the Motor Vehicle Parts Industry. Because of the complex circumstances in which he found himself, Ganz had asked another friend, the engineer Alfred Rüdinger, to hide behind the door to an adjoining room and listen. During the conversation, Goldstein warned Ganz that he was "intensely hated" in some parts of the German car industry and that certain people were out to get him. Goldstein knew about the criminal case Ganz had brought against Paul Ehrhardt, Gustav Röhr, and Bernhard Stoewer, and he stressed that Ehrhardt in particular was an extremely dangerous man to have as a personal enemy. He said he had even heard of a murder plot, in which Ganz would be taken on a test drive, probably in a Mercedes-Benz, and meet with an "accident." Goldstein quoted from the conversation he had overheard: "As long as Ganz is alive, *Motor-Kritik* will not

Datum der Auftragung		Vorgang
16.10.34.	Wirtschaftsspionage.	I.62.00 A.-21-

Vordruck: Pol. 57.

Rear of Frankfurt police registry card for Josef Ganz accusing him of "economic espionage," 1934

ADOLF HITLER
KANZLEI

BERLIN W8
WILHELMSTRASSE 55 II
FERNSPR.: A 2. FLORA 7601

DEN 3. November 1934

Herrn

Philip C u m m i n g s

TAGEBUCH-Nr. 19352/X HP
BEI RÜCKFRAGEN UNBEDINGT ANZUGEBEN

B e r n / Schweiz
~~Bubenbergplatz~~
City Hotel

Sehr geehrter Herr Cummings!

Ich bestätige Ihnen mit bestem Dank den Eingang Ihres an den Führer gerichteten Schreibens und freue mich, daß es Ihnen hier in Deutschland so gut gefallen hat und daß Sie sich davon überzeugen konnten, wie falsch und wie hässlich die im Ausland verbreiteten Greuellügen und -hetzen sind.

Ich spreche Ihnen meine aufrichtige Freude darüber aus, daß Sie dazu beitragen wollen, die Wahrheit über das neue Deutschland zu verbreiten und der Greuel-Propaganda entgegenzutreten.

Mit vorzüglicher Hochachtung!
i. A. [signature]

Letter from the German Chancellery to Philip H. Cummings, 1934 lose its character."[13] Mere days later, the German police officially charged Ganz with "economic sabotage."

Paris was not a safe place for Josef Ganz that week, since both Paul Ehrhardt and Gustav Röhr were visiting the motor show. Shortly after touring the exhibition halls, Ganz left and drove to the Principality of Liechtenstein, where he moved into a room at the Waldhotel in Vaduz. The Waldhotel was a transit point and meeting place for international travelers, and he got to know several people there, among them Philip Harry Cummings, a young man of twenty-seven with wavy auburn hair who worked as a professor of literature in the United States. Now, in the fall of 1934, he was setting out on a world trip, beginning in Europe. Ganz liked the rather naive but friendly young man and met with him at or near the hotel a number of times. Gradually, he told Cummings more about his personal circumstances and the situation in Germany. Cummings had just come from Germany, where he had been treated perfectly decently everywhere, he said, so he found it hard to believe all the negative stories he was reading in the foreign media, but when he traveled to Germany for a second time on November 11, he had a meeting in the train that changed his mind. He was carrying a letter from the German Chancellery, which he showed at the border checkpoint near the German town of Singen am Hohentwiel to help ease the formalities. A fellow passenger, a distinguished looking man, saw the letter and asked Cummings whether perhaps he knew Hitler personally, to which Cummings replied: "Unfortunately not!" He told the man about his positive experiences in Germany, and as the conversation went on he began asking how much truth there was to the stories in the foreign media about German concentration camps. "Hitler has so many enemies that to be able to work freely he's forced to intern these people for observation," the man answered. "It's far better for them to live in concentration camps than to be killed, as has happened in Russia." Cummings said that in Switzerland he had met German refugees who did not dare go back to Germany, and he asked whether they would be allowed to cross the border if they did decide to return. The man, who obviously worked for the German government, told him that the names and passport numbers of

Vaduz, den 2. Dezember 1934.

Ich hatte bei meiner ersten Reise auf dem Flugplatz Mannheim Schwierigkeiten wegen Mitnahme von Devisen über Reiseschecks. Ich erzählte davon meinem Bekannten, dem jungen Herrn Baron L i e b i g, wohnhaft Liberec in der CSR. Dieser hat Verbindungen in Berlin, denen er hiervon berichtete. Darauf erhielt ich mein Geld zurück. Ich bedankte mich hierauf dafür in einem persönlichen Briefe an Hitler, den ich von Arnstadt aus sandte. In diesem Briefe brachte ich zum Ausdruck, dass mir der gute Wille bewiesen wurde, und ich daher nicht mehr das glaube, was man im Auslande alles über Deutschland erzählt.

Ich erhielt hierauf von der Kanzlei des Reichsführers einen sehr höflichem Bestätigungsbrief.

Als ich am 11. November 1934 bei einer neuen Reise nach Deutschland bei Singen - Hohentwiel die deutsche Grenze passierte, habe ich ausser meinem Pass und meiner Brieftasche den offenen Brief der Kanzlei des Reichsführeres so vorgezeigt, dass der Briefkopf sichtbar war, um die Passage mir zu erleichtern. Ich kann mich auf Deutsch, wenn rasch gesprochen wird, nur schwer verständlich machen.

Ich stieg in den Berliner Wagen des Schnellzuges über Stuttgart in ein Abteil zweiter Klasse. In dem Abteil sass ein soigniert aussehender kräftiger Mann mit Brille, der lediglich zwei Handtaschen mit sich führte. Der Mann hatte offenbar gesehen, dass ich dem Grenzbeamten den Brief von Hiler's Kanzlei gezeigt habe.

Der Herr fragte mich in sehr höflichem Tone: "Kennen Sie den Führer?" Ich antwortete: "Leider nicht!" Es entspann sich nun folgendes Gespräch, das ich aus dem Gedächtnis wiedergebe:

Der Herr: "Was denken sie von dem Führer; ich habe bemerkt, dass sie einen Brief von ihm haben."

Ich: "Ich habe ihn gerne, ich sehe, dass alles, was man im Auslande über Deutschland sagt, nicht wahr ist." " Ich sehe viel, was man in englischen und amerikanischen Zeitungen schreibt. Man hat mich in Deutschland immer sehr höflich behandelt."

Dann sprachen wir zunächst über Nebensächliches. Schliesslich fragte ich den Herrn, was wahres daran sei, was man über die Konzentrationslager erzählt, und aus welchem Grund und wer da hineinkommt.

Der Herr: "Hitler hat so viele Feinde, dass er, um eine offene Strasse zur Arbeit zu haben, diese Leute zur Beobachtung konzentrieren muss. Es ist viel besser, diese Leute leben im Konzentrationslager, als, sie werden umgebracht, wie das in Russland gemacht worden ist."

Ich: "Ich habe Deutsche in der Schweiz kennengelernt, die durften nicht in Deutschland bleiben und nach Deutschland zurückkehren."

Der Herr: "Ja, das glaube ich."

Ich Ich: "Wenn diese Leute nach Deutschland zurückwollen, können sie über die Grenze kommen?"

Der Herr: "Ja, aber die Passnummer ist auf einer Liste vermerkt, und auch die Namen dieser Leute, und sie werden bei Grenzübertritt angehalten und vor ein Regierungsgericht gebracht. Dort müssen sie erklären, warum sie so lange im Ausland geblieben sind, auf welche Weise sie Geld aus Deutschland hatten und was sie in der Zeit gegen Deutschland unternommen haben.

Ich: "Werden diese alle in's Konzentrationslager oder in's Gefängnis gebracht?

Der Herr: " In keiner Weise, sie müssen nur befragt werden. Einige könnten allerdings in ein Konzentrationslager verbracht werden." "Haben sie Interesse, einmal eine solche Grenzliste zu sehen?"

Ich: "Natürlich, das interessiert mich, danke Ihnen sehr!"

Der Herr entnahm einer seiner Taschen drei, vier Blätter, die aussahen wie Korrekturfahnen, Folioformat, zweispaltig, ganz frisch bedruckt. Den Kopf mit verschiedenen Erläuterungen, und unterschrieben "Die Kanzlei..." oder so, konnte ich in dem Moment nicht lesen. Da ein Bekannter von mir, Herr Ganz mir einmal Andeutungen über den Bestand einer solchen Liste machte, sah ich rasch unter "G" nach, ob auch der Name von Herrn Ganz auf der Liste steht. Ich sah auf der Liste drei GANZ, der zweite war

Ganz, J. Ing. Frankfurt/M.

Ich bemerkte besonders viele Namen mit der Ortsbezeichnung, Frankfurt/M, Chemnitz, Hannover. Der Herr nahm die Liste rasch wieder an sich. Ich hatte bemerkt, dass einige Namen mit rot durchgestrichen waren.

Ich: "Sind diese, die rot durchgestrichen sind, schon zurückgekommen nach Deutschland?"

Der Herr: (lächelnd): "Nein, wir wissen, dass sie unterdessen gestorben sind, oder sie sind versehentlich auf der Liste."

Ich: "Die Liste ist sehr gross!?"

Der Herr: "Vielleicht zweieinhalb Tausend, und fast alle Juden!" "Und Kommunisten!"

Ich: "Also hat Hitler nicht so viele Feinde."

Hierauf nahm der Herr eine Zeitung und las. Ich stieg in Stuttgart aus, der Herr fuhr weiter.
Ich bin amerik. Staatsbürger, Literaturprofessor. Philip Harry C u m m i n g s,
Z.Z. City-Hotel, B e r n

Philip Cummings

such refugees were on a list. They would be arrested immediately at the border and brought before a government tribunal, then quite possibly sent to concentration camps. The man opened one of his briefcases and pulled out a list of some 2,500 names to show to Cummings, who, with Ganz's story in mind, leafed through it and looked under the letter G. There were three entries for the name Ganz, one of which read: "Ganz, J. Ing. Frankfurt/M."[14] When he returned to Liechtenstein several weeks later, Cummings told Ganz about the encounter. Ganz immediately asked him to set the story down in a statement. In early December, the name "Josef Ganz" also appeared in a book openly published in Germany with around five thousand names of private citizens and businesses suspected of being Jewish.[15]

Diplomatic Passport

Statements by people like Cummings were of vital importance to Josef Ganz. He was traipsing about Europe as a German emigrant with no fixed abode and could not stay anywhere longer than a tourist visa allowed. In mid-December, the immigration police in Liechtenstein discovered that he had spent more than three months at the Waldhotel in Vaduz.[16] An official from the Liechtenstein police visited Ganz and drew his attention to the fact that for a stay of more than three months he needed a proper residence permit. He would need to apply to the Liechtenstein government for one. In his letter of application, Ganz declared that he lived and worked as an engineer in Frankfurt but was staying in Liechtenstein because his life was in danger as a result of "exceptional circumstances in Germany." He added that he was not attempting to find a job in Liechtenstein but was engaged mainly in private work, and he gave the names of various contacts, including Paul Jaray in Luzern, who would be able to

Opposite page: Statement by Philip H. Cummings about his meeting in the train, 1934 vouch for him.[17] In January 1935, Ganz was back in Paris, where through diplomatic contacts he had made an appointment at the consulate of the South American republic of Honduras. The consulate was extremely interested in Ganz's inventions and offered him a position as "technical adviser." As well as a new source of income, the consulate provided him with something of far greater value: a Honduran diplomatic passport.[18] It would allow him to travel around Europe freely. He walked out of the building with a huge sense of relief, burying his new dark-blue passport with its gold lettering deep in the inner pocket of his jacket. Two days later, at the Danish consulate in Paris, he was given a three-month visa for a journey from France via the Soviet Union to Denmark. After several weeks of traveling, he returned to Liechtenstein in February 1935 and again took up residence at the Waldhotel in Vaduz. Mail from Frankfurt was waiting for him there. Georg Ising had forwarded the decision of the Frankfurt court in the case against Ehrhardt, Röhr, and Stoewer. His appeal had been rejected. Although all three had been found guilty, the Law on Immunity made them unimpeachable for their past crimes.[19]

Death Squads

At the Waldhotel in late February 1935, Josef was introduced to one Johann Martin Danner, a young man in his early twenties with slicked-back hair. Like Ganz, Danner was a German refugee who had been wandering around Switzerland and Liechtenstein for almost a year, earning his living by writing articles for the Swiss newspapers. It transpired that he was in fact the former secretary to Reinhard Heydrich, the head of the German security service (the Sicherheitsdienst or SD) and the Gestapo. Danner had worked for both organizations for six months immediately after they were founded in April 1933, when the SD headquarters, with a staff of just six, was located in a twelve-room private house on the Leopoldstrasse in Munich. There, using a card-index system in four steel filing cabinets, they had kept track of the fates of thousands of people. The symbol + on a card meant that, by order of the Gestapo, a person was to be "beaten up" or perhaps thrown out of work. A card with the symbol ++ meant he would be detained and sent to the Dachau concentration camp, and +++ meant the individual concerned was dead.[20] Danner had a great deal of information about, for example, house searches by the Gestapo during which faked documents had supposedly been found, about embezzlement on a grand scale, and about murders committed by the Gestapo or in Dachau. One method was to inject a poisonous protein that caused a heart attack or stroke within about twelve hours and afterwards left no trace.

Because of his categorical refusal to carry out such orders, "which I could not reconcile with my conscience," Danner was arrested in September 1933 and interned at Dachau. There he was kept in solitary confinement for twenty-three days in a cold, dark cell and suffered a nervous collapse as a result. After another five and a half weeks in the hospital wing of the Stadelheim prison in Munich—also an execution site where hundreds of people were guillotined—Danner was freed after SS leader Heinrich Himmler interceded for him. Because of a serious conflict between Himmler and Göring, however, by March 1934 Danner's life was once more in danger, and he fled to Switzerland.[21] Josef Ganz was mesmerized by this story, and he told Danner of his ongoing battle with Paul Ehrhardt and Gustav Röhr, personal friends of Göring. Danner had heard Ehrhardt's name before, and he told Ganz that Ehrhardt was known to the SD as a "technical and political professional agent" who had served in the counterespionage department of the Air Ministry under Göring. He warned Ganz that Ehrhardt went under the false name of Schröder and had carried out a number of secret missions in Switzerland.[22] Danner even claimed the Gestapo had hired a hitman to kill Ganz and said he could expect no protection at all from the German police.[23] He had incredible stories to tell about SS death squads that had been carrying out liquidations since the spring of 1933 well beyond the German border—Liechtenstein, Austria, and Switzerland included. One example was the killing of the German engineer Georg Emil Bell, the former secretary to SA leader Ernst Röhm, who had been riddled with bullets in Austria in April 1933 by a five-man hit squad that fired at him from a Mercedes-Benz. Danner was so well informed that he was even able to describe how the men had afterwards driven straight through the barrier at a Swiss border post, damaging a headlamp and the radiator. The engine had overheated as a consequence, and repairs to the car had cost some 800 Reichsmark.[24]

Ganz was appalled. He now understood the stark fact that he was not safe in Liechtenstein. Danner typed on Ganz's typewriter several statements they could both use to support their stories as part of their efforts to obtain residence permits for Liechtenstein. It was not to be. At the end of March, Danner received official notification that he must leave the country within eight days. His future was extremely uncertain. He would try to cross Austria to reach friends in Czechoslovakia. A little while later, Josef Ganz received the shocking news that Danner had been murdered by the Gestapo. He feared the same fate awaited him. The terror and homicide of Nazi Germany did not stop at national borders, and he might fall prey to a death squad at any time.

Soon afterwards, he left Liechtenstein and traveled to Switzerland, where he visited the Geneva motor show in March 1935. At the Tatra stand he spoke to the factory worker Zdenko Kochanowsky, who said he had originally been asked by Tatra's director Petrina to drive through Frankfurt on his way from Paris to Geneva to pick up Paul Ehrhardt. That plan was dropped at the last moment, and Ehrhardt had not come to Geneva—Ganz suspected Ehrhardt was aware of the statements he had made to the Swiss police about him and had decided it was too risky to travel.[25] Ganz heard yet more bad news when he rang Christl-Marie Schustel, a German pilot and aviation pioneer he had met at the Waldhotel in Vaduz. Schustel, much like Ganz, had fled Germany in November 1933 after she was forced to resign as editor-in-chief of the magazine *Deutschen Flugillustrierten*, which she had founded, because of her refusal to become a member of the Nazi Party.[26] She now told him that shortly after he left Vaduz, a German police detective called Kaufman Cane had arrived at the Waldhotel and tried to wheedle out of her the address where Josef Ganz was staying.[27] The hunt was on for the Jewish émigré.

Josef Ganz's Tatra waits in the back of a line of cars (on the left) in front of the German-Swiss border, 1934. This was the holiday that ended up being the beginning of Ganz's final exile.

Top: Brochure for the new model of the Standard Superior proclaiming it to be "the fastest and cheapest German Volkswagen" with space for a typical German family, 1933.
Bottom: The last brochure for the Standard Superior just showed a photograph of the car bearing the caption: "The most reliable, comfortable, and economical small car," 1934.
The Nazis forced all companies but VW to stop using the term Volkswagen in their advertising.

Josef Ganz relaxing at a Swiss mountain lake, unaware of the latest catastrophic events happening in Germany, 1934. His holiday to Switzerland would save him from arrest and possible assassination by the Gestapo on the Night of the Long Knives.

Above and three images on next page: Paul Jaray behind the wheel of a Mercedes-Benz prototype with streamlined bodywork according to his design, driving to Mount Pilatus, Luzern, 1934. Josef Ganz visited Jaray while on holiday in Switzerland. The teardrop shape of the Mercedes-Benz prototype closely resembled a design Paul Jaray made for the Bungartz Butz, developed according to the patents of Josef Ganz, 1934.

Above and left: A Volkswagen design by Paul Jaray for the chassis of the Bungartz Butz designed by Josef Ganz, 1934.

VERTEX
IDEAL
SCINTILLA
START
VERTEX
Esso

Top previous page: The start of the Swiss Klausen Rennen, which Josef Ganz attended after being unable to return from his holidays in Switzerland, 1934.
Bottom previous page: Racing driver Rudolf Caracciola is about to board the Mercedes-Benz W25 at the Klausen Rennen, 1934.
Above: Ferdinand Porsche (with hat) next to his Auto Union Type A racing car, just minutes before the Klausen Rennen begins, 1934. Behind the wheel: renowned racing driver Hans Stuck.

Jakob Werlin from Daimler-Benz (left of woman) next to Luigi Fagioli (in Pirelli overalls) after the famous Italian racing driver had won him the Coppa Acerbo Grand Prix in Pescara (Italy) steering a Mercedes-Benz W25, 1934.

Top: Josef Ganz spoke at the Klausen Rennen to Ferdinand Porsche, who had just signed the contract to develop the Volkswagen for Hitler, 1934.
Bottom: Jakob Werlin (left), a protégé of Hitler, was responsible for the official publication ban that ended Ganz's German career in 1934. This effort aimed at erasing Ganz's name from all Nazi records that could have linked him to the design of the VW Beetle.

Hitler addresses the crowds at the NSDAP Parteitag in Nuremberg, September 1934. While in Nazi Germany all attention was focused on this event, Josef Ganz made a hasty trip to Frankfurt to retreive his records. The Gestapo had come after him, and he was no longer safe on German soil. But Ganz refused to go into exile without his precious archive still stored at his Frankfurt office.

Josef Midinet, an old family friend of Josef Ganz, next to the Röhr Junior Limousine after it broke down in Germany during the secret trip to retrieve Ganz's documents, 1934.

Josef Ganz has his last tea ever in the garden of his Frankfurt home and office, just hours before fleeing the country, taking with him all the designs his car could carry. He is joined by the wife of his friend Georg Ising (his colleague at *Motor Kritik*), Gustel (left), 1934.

The Röhr Junior Limousine, filled with documents, as Josef Ganz and his friend Josef Midinet try to rescue Ganz's vast archive, just moments before they leave Frankfurt for the Swiss border. In the meantime, Hitler and his NSDAP celebrate their Parteitag in Nuremberg. This event, Ganz and Midinet reckoned, will divert attention from their secret and illegal endeavor, September 1934.

Josef Midinet on the steps of a hotel in the Black Forest, the Röhr Junior Limousine parked out front. He and Ganz are on their way to the Swiss border. Ganz waited around the corner from where he took this photo, while Midinet checked the hotel for Nazis.

Right: Three men who supported Josef Ganz during his exile, from top to bottom: streamlining pioneer Paul Jaray, ex-Gestapo officer Johann Martin Danner, and American literary professor Philip Harry Cummings.

PERSONAS QUE LE ACOMPAÑAN

Firma del portador
Dipl. Ing. Joseph Ganz

REPÚBLICA DE HONDURAS

Secretaría
de Relaciones Exteriores

La Legación de la República de Honduras en París, Francia, concede libre y franco pasaporte al Sr. Ingeniero Don Josef Ganz, Consejero Técnico del Consulado General de Honduras en Berlín, Alemania para que pueda viajar por cualquier país

París, 19 de Enero de 1935

Josef Ganz was able to get a position as "technical adviser" to the Honduran government and received a Honduran diplomatic passport, which allowed him to travel around Europe freely, 1935.

Josef Ganz after settling in Switzerland, around 1935.

Josef Ganz (left) and a Hungarian uncle with the original May Bug prototype, around 1935. Ganz had the May Bug transported to Switzerland and demonstrated the car for the Swiss Ministry of Employment.

Top: In 1936, Josef Ganz moved to this apartment building in Zürich.
Middle: The home office of Josef Ganz where he would work through the night, around 1938.
Bottom: At his workbench, Josef Ganz constructed models of the Swiss Volkswagen, around 1938.

Josef Ganz comfortable on the roof of a Standard Superior with Swiss registration, Zürich, around 1935. Demonstrations of this car and the May Bug for the Swiss Ministry of Employment helped Ganz to secure the Swiss Volkswagen project.

ZH
17584
CH

Top: The prototype backbone chassis of the Swiss Volkswagen was based on the design of the Standard Superior but fitted with a single-cylinder engine, around 1936. In the background is Ganz's private car, a Packard 645 Phaeton.
Bottom: Madeleine Paqué, the girlfriend of Josef Ganz, joins engineer Jean Grünbaum for a test drive of a prototype chassis of the Swiss Volkswagen, around 1939.

Top: A factory worker attaches the bulbous nose section on one of the prototypes of the Swiss Volkswagen, around 1937.

Bottom: The body of the Swiss Volkswagen was fully constructed out of sheet metal.

Top: Josef Ganz drives one of the prototypes of the Swiss Volkswagen from the driveway of his apartment building into the street, Zürich, around 1937.
Bottom: The Swiss Volkswagen prototypes were severely tested in open terrain, such as forests and fields, around 1937.

Above: Test driving a prototype of the Swiss Volkswagen through the streets of Zürich, where bicycles were the most common form of transport, around 1937.
Left: Model of a Swiss Volkswagen, with a newly designed nose section with built-in headlights behind grilles for protection, around 1938.

Top: Swiss soldiers carry the Swiss Volkswagen prototype across the street, 1938.
Bottom: Josef Ganz is steering the aluminium-bodied Swiss Volkswagen prototype, which he nicknamed the Silverfish, around 1938. This model has a new nose design, with headlights behind grilles.

Top: The Swiss Volkswagen prototype attracts crowds in the streets of Zürich, around 1938.
Bottom: Josef Ganz acquired special permission to keep test driving his Swiss Volkswagen prototype during the war, 1941.

Top: Ferry Porsche is steering the second Volkswagen prototype, the V2 convertible, 1936.
Middle left and right: The first Volkswagen prototype, the V1, 1936.
Bottom left and right: The Volkswagen prototype V30, 1937. Daimler-Benz built a limited series of just thirty of these cars, which were extensively tested.

Top: Ferdinand Porsche (left) presents Adolf Hitler with a scale model of the new Volkswagen for his forty-ninth birthday, 1938. On the right stands Jakob Werlin and in the middle background Adolf Hühnlein, leader of the NSKK.
Bottom: Pre-production model of the Volkswagen or KdF-Wagen, 1938.

Josef Ganz drove the Silverfish to St. Moritz to attend the Olympic Games, where the Swiss team won gold and silver medals using a bobsled fitted with an innovative suspension system for skis developed by Ganz, 1948.

Top: Ferdinand Porsche, after his release from prison, with a Volkswagen in the Swiss Alps, 1950. That year, Porsche and Josef Ganz met each other for the last time in Geneva. Porsche died the following year.
Bottom: Heinrich Nordhoff (second from left), director of Volkswagen, next to the one millionth Volkswagen Beetle produced in the factory in Wolfsburg, 1955.

Top: A Julien MM5, based on the Swiss Volkswagen design by Josef Ganz, at the factory in France, around 1950. In the background, the larger model Julien MM7.
Bottom: Final inspection of the Julien MM5 (left) and MM7 models at the factory in France, around 1950.
Right: Josef Ganz on the ship to Australia, 1951.

Top: Heta Jacobson, Josef Ganz's new girlfriend, with whom he lived in Australia, around 1955.
Bottom: Josef Ganz with his Australian naturalization certificate, 1962.

Top and bottom left: A prototype of the pickup, convertible prototype of the Standard Ten at the Standard Motor Company (Australia) Pty Ltd, where Josef Ganz had found work as an engineer, around 1953.

HOW I INVENTED THE VOLKSWAGEN

The car reached the prototype stage, but was not produced. Porsche left Mercedes the next year and, early in 1929, started work at Steyr, where he designed a straightforward but practical 2-litre car and a 5.3-litre model called the Austria, which was the first of his cars to have swing axles at the rear.

At the same time I was busy working on my ideas for a revolutionary Volkswagen. In 1929 I prepared a written submission, setting out in full my ideas for the production of my car for the technical director of the giant Zundapp motor-cycle company, Herr Wittig. He decided not to go ahead with it at that time, and when Zundapp were prepared to produce a small car in 1931 I was already tied up with the Adler Company.

In 1930 the Ardie Motor-cycle Company produced an experimental prototype to my design. The idea of this car was to prove my point that a small car be made which would be comfortable and handle well over all kinds

BELOW: The cover of the September 1930 Motor Kritik showed Ganz behind the wheel of his first experimental car.

ABOVE: The first brochure for the Standard "Superior."
BELOW: Hitler made the company drop the name Volkswagen in 1934 when he decided to use it for political gain.

of roads. It did this with a vengeance and even today it would be hard to find a car which rode as well.

The economic situation in Germany at the time was such that not many companies were prepared to start new ventures. Ardie did not want to build the car and, in 1931, the Adler company took up the design and built another, more powerful prototype. They also decided against going into [illegible] production.

Then, in 1932, the Standard Motor-cycle Company decided to make the move which Germany's other motor-cycle manufacturers were dreaming of, but not brave enough to make.

The car which Standard produced created quite a stir when it was first released. An American newspaper, the Detroit News, commented on its "Roomy dimensions, low pitch, low centre of gravity, good riding and road-holding qualities."

As you can see from the illustrations, the car was streamlined, had a rear engine (a 500 cc two-stroke), a platform backbone chassis, independent suspension by swing axles at the front and pendular axles at the rear, and it looked, I must admit, like a bug-eyed beetle.

On February 11, 1933, the newly elected Chancellor of Germany carried out his first official duty. Adolf Hitler, the painter who became the leader of the National Socialist Party, opened the annual Berlin Motor Show.

Before the show officially opened, Reichsmarshal Goering strutted through the exhibits, stopping here and there when a particular model caught his attention.

Eventually he arrived at my strange looking car. It was called the Standard Superior and its brochures proclaimed it to be a Deutschen Volkswagen (German people's

ABOVE: The prototype built by the Adler motor-cycle company in 1931.
BELOW LEFT: This is how VW looked when it went into production. Prototypes had more protruding headlights.
BOTTOM LEFT: The Standard Superior, which was released in 1932 had many similarities to the VW, which was released six years later.
BELOW: Josef Ganz. He now lives in retirement in Melbourne.

I CREATED the Volkswagen. I coined the name and created a car with rear engine, backbone platform chassis and swing axle suspension at a time when Dr. Ferdinand Porsche, as technical director of Mercedes-Benz, was scrapping all that company's rear-engined prototypes because he thought the system impractical.

Dr. Porsche, whom I respect as a fine engineer, only completed the final engineering for the VW's production. I am sure, if he were alive today, he would agree with what I'm about to tell you.

My first work on the design of what was to become the Volkswagen started in 1923 when I was a consultant automotive and radio engineer and motoring writer for Motor Kritik magazine. I wrote articles criticising the German motor industry and its way of building cars that were old fashioned, heavy, unstreamlined and expensive both to build and run.

I wanted to see a car which was the exact opposite and which the average working man could own. In June 1924 I first presented the public with my ideas for such a car.

It has swing axles and the engine and transmission in one unit just in front of the rear axle.

In a series of illustrations I explained why the car would handle well with its swing axle and how it could be streamlined.

I continued to write articles pointing out the faults of current car designs, and shortly afterwards wrote an article headed "No More Cars Without Swing Axles".

The motor industry did not take kindly to my criticisms and although many engineers agreed with me and became my friends, others became my bitter enemies. Just how bitter I wasn't to realise until Hitler came into power.

In 1928 I became editor of Motor Kritik and immediately started a campaign for what I called the Deutschen Volkswagen (German people's car). I campaigned for a special low registration fee for small cars, which the government eventually granted.

In 1927 Ferdinand Porsche, who was technical director of Mercedes-Benz at the time, designed his first small car. The car was perfectly straight-forward, with the engine in the front and a live rear axle — he had already discarded the company's rear-engined prototypes.

In *Australian Motor Sports & Automobiles* Josef Ganz tried one last time to set the record straight, 1965. *Left:* The Holden FJ of Josef Ganz, built by GM Holden, where Ganz found work at the Engineering Department, parked (on the left) in St. Kilda, around 1955.

Josef Ganz, Melbourne, around 1965.

Röhr Joins Daimler-Benz

Josef Ganz had been trekking around Europe for more than seven months when, in March 1935, he arrived at Montreux in Switzerland. There he took a room for a while at the Hotel d'Angleterre on the shores of Lake Geneva and resumed his correspondence with friends and colleagues in Germany. Georg Ising in Frankfurt sent him a parcel containing the latest issues of *Motor-Kritik* and letters that had arrived for him, including one from the president of the Reich Chamber of Literature in Berlin, who informed him that his membership in the National Association of German Authors would be terminated on April 1, 1935. The letter pointed to a regulation that had been put in place according to "the will of the Führer and Reichschancellor." It meant that "only persons who belong to the German people not only as citizens but through deep connections of character and blood are suitable to carry out highly important intellectually and culturally creative work in Germany for the life and future of the German people." Ganz, because of his "quality of being a non-Aryan," was not in a position to "feel and recognize such a duty."[28] This official step, disqualifying him as a journalist, was rather belated. He had already been banned from publishing in Germany for over a year, and he could not even enter the country without facing immediate arrest. His position was extremely weak, and the case against Paul Ehrhardt, Gustav Röhr, and Bernhard Stoewer had come to a complete standstill.

He read in *Motor-Kritik* that the contract between Gustav Röhr and Adler would not be renewed and that Röhr was about to be appointed technical director at Daimler-Benz, succeeding Hans Nibel, who had died in Stuttgart in late November 1934. Joseph Dauben and all the rest of Gustav Röhr's technical staff would follow him there. Ganz had met a lawyer for Daimler-Benz in Vaduz in early 1935 who confirmed to him that his dismissal in late 1933 had been decided upon not by the directors but "as a result of external pressure."[29] Ganz told his lawyer Theodor Hartherz in Frankfurt the latest news about Röhr and described the shocking things Johann Martin Danner had revealed about Ehrhardt.[30]

Writing back, Hartherz said that the reports on Ehrhardt were "not new" to him.[31] Hartherz was an ex-member of the Nazi Party, and he had heard that Ehrhardt enjoyed protection in return for "valuable services to the party." As Josef Ganz would later discover, those services may have had something to do with the starting of the fire at the Reichstag, the event that had accelerated Germany's descent into absolute dictatorship. From Werner Wolfgang Knoeckel—the former editor-in-chief of the German magazine *Südwestdeutsche Rundfunk-Zeitung*, which had the same publisher as *Motor-Kritik*—Ganz heard that Göring had composed a written declaration forbidding the German authorities to interfere in the affairs of Paul Ehrhardt. Knoeckel, who was now working as a civil servant in radio broadcasting, had witnessed how Ehrhardt was able to use the declaration to get himself a temporary post as a radio reporter.[32] One scrap of information about Paul Ehrhardt after another filtered through to Ganz from Germany—his personal enemy was utterly untouchable.

Criticism of the Third Reich

It had become terrifyingly clear to Josef Ganz that the Third Reich was keeping a close eye on German emigrants. In July 1935, an extremely critical article appeared in the Swiss newspaper *Berner Tagwacht* about the expensive racing cars built by Auto Union and Mercedes-Benz, propaganda machines for the Third Reich that were achieving one victory after another. "Germany is, however, one should be aware, a poor country," the anonymous author wrote, "since it cannot pay its debts! But in 1933 five million Reichsmark was made available to Auto Union and another five million to Mercedes-Benz." The racing teams for the two companies traveled nonstop around Europe with the cars they had built, and they could afford the very best drivers. The author described the famous racing drivers Rudolf Caracciola, Achille Varzi, Luigi Fagioli, and Hans Stuck as "men with courage and nerve, who do however cost money." Their work involved "deadly danger, pain, and blood, in return for money, a great deal of money... Good advertising is always costly."

The propaganda machine rolled on, but "about the Volkswagen one hears nothing at all any more." The Third Reich, the writer of the article said, was a facade behind which millions of Reichsmark were disappearing into the pockets of party bigwigs, "a Roman circus under the sign of the swastika."[33]

The article immediately came to the attention of Nazi Party officials in Germany and among the people it was shown to were Adolf Hühnlein of the NSKK[34] and Jakob Werlin of Daimler-Benz. Werlin contacted the director of Mercedes-Benz in Zürich, Erich Friedrich Muff, and asked him to provide the name of the author of the article, saying he must deal with the whole matter "personally and in the strictest confidence."[35] Muff made cautious inquiries, but the editors of the *Berner Tagwacht* would tell him nothing. He did remember that several months earlier he had been visited at Mercedes-Benz by a German émigré who was hoping to be given a position at the company. The man had been rejected immediately and his name had not been registered, but the staff recalled that he had been a redhead and that his name, which began with B, was the same as that of a town in Germany. Muff reported to Werlin that it had perhaps been Rolf Bielefeld, formerly *Motor-Kritik*'s motorcycle reporter, "although there is no certainty on the matter."[36] Muff's investigation eventually ran into the sand. He was able to conclude only that the author was "a journalist living abroad."[37] That journalist may well have been Josef Ganz.

CHAPTER FIFTEEN

SWISS VOLKSWAGEN

Ganz left the Hotel d'Angleterre in Montreux that summer, and at the end of August 1935 he found a permanent place to live at In der Hub 16 in Zürich. From there he tried to resume some of his usual activities, such as the defense of his international patent rights. At the Geneva motor show, he had noticed that on its new model, called the Baby, the Czechoslovakian car manufacturer Praga had used swing axles of a type on which he held the patent. In order to take judicial steps against Praga, Ganz contacted a Swiss lawyer, Dr. Ferdinand Fuchs of Zürich. He won the case. Praga paid damages of twenty thousand Swiss francs for the unauthorized use of Ganz's design.[1] During the negotiations, Fuchs heard a good deal about Josef Ganz's background, inventions, and patents. He was greatly interested in the Volkswagen, and with the help of Georg Ising of *Motor-Kritik* he had the Maikäfer prototype transported to Switzerland from Frankfurt, along with Ganz's Tatra 11. After an extensive test drive in the May Bug, Fuchs told Ganz that he believed there must surely be a large market for such a car and that starting up a Swiss car industry would be a way to create thousands of jobs. With Fuchs as mediator, Josef Ganz came into contact with the Ministry of Employment in Zürich and was given a chance to demonstrate both the Maikäfer and a Standard Superior with Swiss license plates. The men from the ministry responded enthusiastically, saying the Maikäfer had "astonishingly good road holding and an ability to cope with open terrain," and they resolved to investigate opportunities for producing such a car in Switzerland.[2] There was also some interest in the possibility of a military version, and before the summer was out Ganz had demonstrated the May Bug to representatives of the Swiss Ministry of Defense.

Gestapo in Switzerland

Although in Switzerland Ganz had found a temporary port in a storm, the Gestapo turned out to be just as active at espionage there as it had been in Liechtenstein. In late October 1935, he received an unexpected visit from two Swiss police detectives, Hüni and Frei, who told him that "a certain Herr Ehrhardt" from Germany had accused him of being "a crook" and that they had been ordered to investigate the case. Ganz did his best to explain the complex espionage affair surrounding Ehrhardt and gave them a dossier with information and witness statements concerning Ehrhardt and his accomplice Gustav Röhr.[3] One of the most recent reports came from Dipl.-Ing. Amthor, a friend of Ganz who lived across the street from Ehrhardt in Frankfurt. He testified that several murders had been committed in Ehrhardt's house and the bodies driven away "in a silver-gray car with the back seat covered." By this point, Ganz had heard that Ehrhardt might have been one of the arsonists involved in the Reichstag fire. He ended his account in a letter to one of the detectives by saying: "Ehrhardt knows that I've seen through him and is therefore my mortal enemy."[4] For the time being at least, his revelatory statements about Ehrhardt were enough to put a stop to the police investigation.

Representatives of the Ministry of Employment, meanwhile, had asked three independent Swiss automobile experts to evaluate Ganz's Volkswagen designs: Dipl.-Ing. Edwin Oetiker, Dipl.-Ing. MAC Troesch, and Ad Brüderlin. They came to the unanimous conclusion that the car was "extremely good" and admirably suitable for serial production in Switzerland. Brüderlin's preliminary calculations suggested that 54,000

Swiss francs would have to be set aside for the development, building, and testing of three prototypes.[5] The lawyer and engineer Fritz Isler looked at Ganz's patents and concluded that "technically nothing stands in the way of further pursuance of this matter."[6] A committee of technicians was appointed to take charge of the "Swiss Volkswagen" project, consisting of Josef Ganz, Troesch, Brüderlin, and the engineer P. Huber of the Ministry of Employment. The future was looking good. Everything pointed to Josef Ganz being able to fulfill his dream of creating a Volkswagen in Switzerland, in parallel with that of Ferdinand Porsche in Germany.

In his speech at the IAMA in 1935, Adolf Hitler had for the first time openly named Ferdinand Porsche as the designer of the German Volkswagen, saying the initial prototypes would be tested in the summer. His timetable was completely unrealistic. The first two Porsche prototypes, the V1 and the V2, were tested in early 1936, around the time of the next IAMA in Berlin. In early February 1936, at a crucial meeting about the development of the Volkswagen, Porsche once more indicated that it was virtually impossible to develop a proper car with four seats for the RM 1,000 retail price Hitler insisted upon, "only a two-seater, perhaps."[7] To keep the price down to RM 990, more and more costs were excluded from the preliminary calculations, such as capital investment, write-downs, profit-taking, interest payments, and retailing expenses.[8] Familiar with Porsche's work on the Volkswagen, editor-in-chief Georg Ising sent a postcard to Josef Ganz that month on behalf of the editorial team. It read: "In response to the 1936 motor show, the thoughts of the undersigned turn to the auto pioneer and champion of the Volkswagen, whose work is being made a reality at this time."[9]

Gutbrod Visits Porsche

By February 1936, Ganz had moved to a rather larger apartment at Witikonerstrasse 275 in a suburb of Zürich, where he could live with his girlfriend Madeleine Paqué. After the death of her mother in Frankfurt, she too had left her life in Germany behind and emigrated to Switzerland to be with Josef. In late 1937, she was followed by her sister Ute Klüpfel and Ute's son Dieter, who also intended to start a new life in Switzerland and moved into the same apartment. Seven-year-old Dieter's pet names for Josef and Madeleine were Uncle Seppl and Auntie Lö. They were not to be granted a quiet life, however. When his allegations against Ganz came to nothing, Paul Ehrhardt refused to let things rest. He submitted new charges, which led to Ganz's arrest in March 1936 in Lausanne, where he had arranged to meet with director Hans Ledwinka of Tatra in the run-up to the Geneva motor show.[10] Ganz once again made a detailed statement, and he was supported by Ledwinka. He was released later that day. At his next meeting with Ledwinka, Ganz tried to convince him, based on patents, technical articles, and witness statements both that the court case Tatra was still pursuing against him was groundless and that Tatra had taken on board a dodgy character in the person of Paul Ehrhardt.[11] Although Ganz continued to have a good relationship with Ledwinka, the directors of Tatra went ahead with their case against him.

As part of the court's attempt to reach a verdict, the independent lawyer Dr. C. Wiegand was engaged at the beginning of May. In his extensive report, he concluded that in drawing up his patent and applying it in the Standard Superior, Ganz had used Tatra's patent "neither fully, nor in any essential part."[12] Wiegand's report led to a decision by the Patent Office in Berlin in late June 1936, which stated it was "not possible to come to the conclusion that there is any common ground between the patents DRP 587409 and DRP 469644," since "the mounting of the engine described by each patent is quite different."[13] Tatra's position was now extremely weak, but the case had led Wilhelm Gutbrod at Standard to end production of the Standard Superior in May 1935.[14] He would later explain that he had done so purely for fear of "an unfavorable verdict." The Standard Superior had proven an "excellent" car, and years after production ceased it could still be seen driving around "satisfactorily," even after clocking up 200,000 miles.[15] In the summer of 1936, the German journalist Walter Krumnow published a glowing test report on a Standard Superior from its first year of production in motorcycle magazine *Das Motorrad*. He wondered why this car, with its "excellent driving qualities," was no longer being produced. A year earlier, Krumnow had made a lengthy tour of Switzerland in it. During the trip, people had asked him again and again "whether this was the new Volkswagen," the car that was causing such a stir in Germany.[16]

By then, Hitler had not even seen the Volkswagen prototypes. In the summer of 1936, he summoned Jakob Werlin to the Reichs Chancellery for a personal update. Hitler's vocal chords had just been operated on, and he wrote down for Werlin on a notepad, "I can not speak." He proceeded with a number of detailed questions about the engine configuration, power output, and air-cooling, as well as the weight, maximum speed, and fuel consumption of the prototypes. Porsche, however, would not hand over the prototypes to the RDA for testing until October.[17]

Production of the Standard Superior having ended, in September 1936 Tatra offered to drop its court case. In response, Wilhelm Gutbrod of Standard contacted chief engineer Otto Schirz of the RDA—the organization entrusted by Hitler with the development of the Volkswagen—in the knowledge that Tatra intended to take legal action against other "similar designs" in Germany besides the Standard Superior, "especially those constructed by Herr Porsche."[18] On October 5, 1936, Gutbrod visited Porsche's design bureau to discuss ways of neutralizing Tatra's patent, so that it would no longer present a threat.[19] In light of the significant differences between the design of the Volkswagen and the Tatra patent, Porsche eventually decided "to take no further steps in connection with this affair."[20] That October, however, the court case took a bizarre turn when the Standard Fahrzeugfabrik and Wilhelm Gutbrod were fined RM 10,000 by a higher court as the result of an appeal. In the same decision, Josef Ganz was fined RM 3,000.[21] Ganz contacted the lawyer Fritz Isler who, after studying the licenses, came to the same conclusion as Wiegand: "A violation of patent DRP 469644 with a vehicle built according to patent DRP 587409 is completely inconceivable."[22] Josef Ganz suspected that Tatra had managed to influence the outcome of the case through Paul Ehrhardt's Gestapo contacts at the Patent Office. He did not intend to let the matter rest there.

Swiss Volkswagen

The court battle with Tatra did not present any immediate obstacle to the development of the Swiss Volkswagen, since there was no Swiss version of the Tatra patent. On October 28, 1936, a contract was signed between Ferdinand Fuchs and the Ministry of Employment in Zürich for the development of a Volkswagen according to Ganz's Swiss patents. Back in the spring, however, Fuchs had been warned by a good friend of Ganz, Prof. Dr. Ing. Georg Schlesinger—a former member of the staff at the Technical College in Berlin—that "the German patents of Herr G. are at risk as long as they remain in his hands." Schlesinger had written to Fuchs, punning on Ganz's name (Ganzen being German for geese):

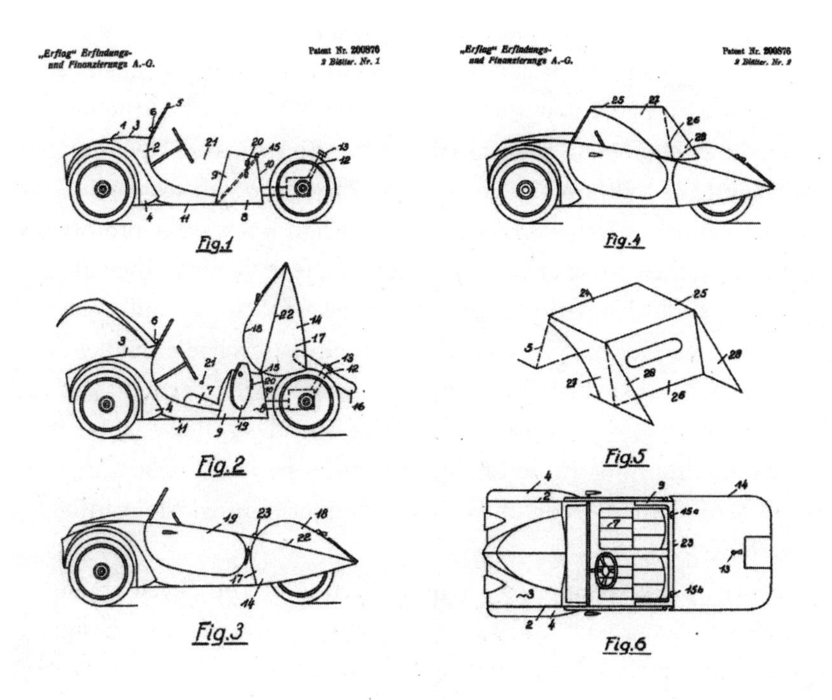

Drawing from CH patent #200876 for the Swiss Volkswagen, filed by ERFIAG in 1937

> I do not think that it will be appreciated in Germany if the building of the so-called Volkswagen is obstructed by the patents of a Jewish émigré, even if he be as well liked as a result of them as Herr G. I know a great many leading automobilists—and they are playing first fiddle in Germany at the moment—who will storm out this very day at the mere mention of his name. Geese generally gabble, but when the gabbling starts to resemble a sensation-seeking noise, people will try to wring the bird's neck.[23]

To protect his intellectual property rights, Ganz decided to sell his most important patents to a new company, ERFIAG (Erfindungs- und Finanzierungs AG), set up in June 1937 with money from external financiers and with Ferdinand Fuchs as director. As a result of the ongoing Swiss Volkswagen project, Ganz was given a Swiss residence permit in 1937, valid for one year. He would be left in peace as far as that went until 1938 at least, and if the project was successful, he would be able to stay longer. He also had a new source of income. With some of the money he bought an eight-year-old Packard 645 Phaeton—a large, open-topped American car—that was the precise opposite of the Volkswagen he was aiming to create. After several months of design work, three prototypes were built in the course of 1937 by Motosacoche in Geneva. They initially used a Standard Superior chassis and its original two-cylinder engine, imported from Germany. A second prototype chassis had a smaller one-cylinder, two stroke, 0.35-liter engine. The car weighed around 300 kilograms empty (661 pounds), and its three-speed gearbox enabled it to reach a maximum of around fifty-three miles per hour. Consumption was an astonishingly economical seventy-eight to ninety-four miles to the gallon.

In 1937, ERFIAG applied for new patents for the development of the Swiss Volkswagen[24] and a military vehicle with a chassis designed by Ganz that included a mounting for a machine gun.[25] The patented Volkswagen design, like the Standard Superior, had a backbone chassis with the engine and gearbox mounted immediately in front of the driven rear swing axles, and independent suspension. It was a streamlined convertible with two seats, looking much like the original Ganz design of 1923. The entire back shell covering the rear wheels, engine space, and spare wheel hinged up. Ganz had deliberately opted for an open two-seater because it was the cheapest to build. He believed there was a market for a larger, closed, four-seater vehicle, but it would be 30 percent more expensive, which would lose it 90 percent of its potential buyers.[26]

Test Drives

Josef Ganz thoroughly enjoyed carrying out comprehensive test drives in the Swiss Volkswagen. He regularly drove at full throttle across the sloping fields near his house on the edge of Zürich, scaring the living daylights out of local farmers. Like the Maikäfer and the Standard Superior, the Swiss Volkswagen prototype attracted crowds of people as soon as it appeared on the streets of the town. In the fall of 1937, the prototypes, having been spotted in Zürich on several occasions, were presented to the press. *Automobil-Revue* wrote that the "Swiss Volkswagen," designed by Josef Ganz, featured "above average road holding and driving comfort."[27] An editor at the magazine *Touring*, which had tested the car alongside Ganz, wrote that the Volkswagen "had been given an extremely simple finish, but by no means at the expense of its comfort, capacities, or robustness." He wrote that the car "was more spacious than almost any other small car on the market, had a smooth gearshift, and could be steered with one finger on even the worst roads." He described its handling as "outstanding" and was very pleased by the way it looked.[28] A competition was held to devise a name for the car, and dozens of people responded with suggestions such as Gotthard, Rapide,

Logo for the KdF-Wagen, 1938

Ganz Autoli, Ganzette, Ganzoto, Ginganz, Voga (Volkswagen Ganz), and Svaga (Suisse veritable auto Ganz). In *Motor-Kritik* Georg Ising published a short, businesslike article with an accompanying photograph of the "small Swiss car design."[29] Nazi censorship prevented him from naming its designer.

In October 1938, a thorough six-day test drive was carried out with one of the prototypes, covering a distance of more than seven hundred miles all over Switzerland, from Zürich to Bern, then on via Lausanne to Geneva, through the southwestern part of the country, and back to Zürich. The comprehensive report concluded that the test drive had gone "splendidly," aside from a few minor problems. Nothing seemed to stand in the way of production.

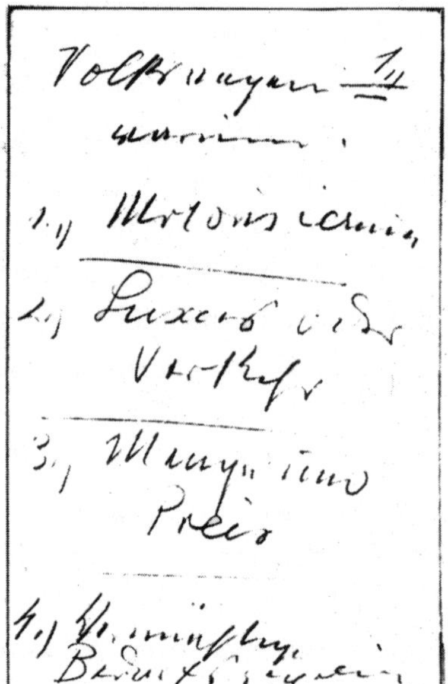

That same year, Ganz developed a new prototype, giving it aluminum bodywork to save even more weight. The design was adjusted slightly too. The headlamps were now mounted behind grilles to protect the glass against stones thrown up by the Swiss mountain roads, and the hinged front windshield could be folded down. Ganz gave the shiny polished-aluminum version the nickname Silverfish.

For serial production of the Swiss Volkswagen, approaches were made to Rapid Motormäher AG, a producer of motorized agricultural machinery in Zürich. It would first produce twenty to forty pre-serial cars and then gradually increase production to five hundred or a thousand vehicles annually—in stark contrast to the mass production of a million and a half Volkswagens per year planned in Hitler's Germany. To increase the number of production models available, Rapid would research the possibility of producing a delivery-van version, a military version, and a passenger car with four seats and a weight of 400-500 kilograms (882-1102 pounds), all based on the Swiss Volkswagen design.[30] Meanwhile, the Swiss development consortium was keeping close track of events in Germany. In 1937, Jakob Werlin had ordered his company, Daimler-Benz, to build a series of thirty new Volkswagen V30 prototypes, which were tested throughout the summer and winter under all kinds of conditions over 1.8 million miles of roads by a group of SS soldiers. Further development resulted in the VW38 pre-production cars. On May 26, 1938, at a bombastic ceremony in Fallersleben (later renamed Wolfsburg), Adolf Hitler laid the first stone of the largest car factory in Europe, where the Volkswagen, still at the initial pre-serial

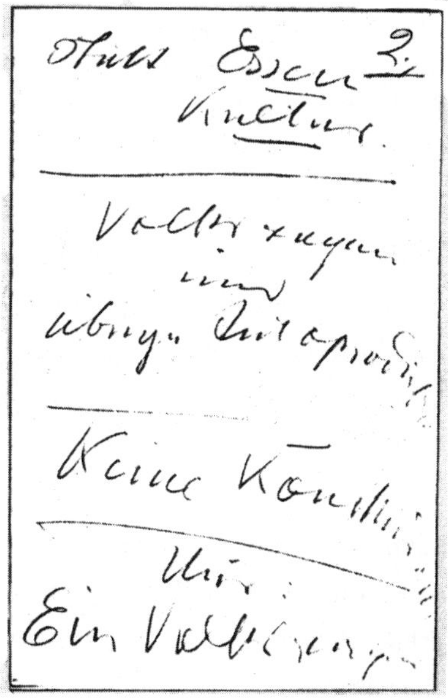

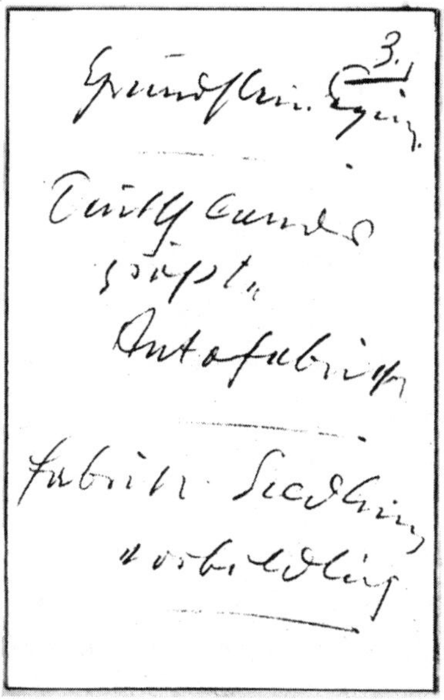

Opposite page and above: First three notes, out of a set of five, created by Hitler in preparation for his speech at the opening of the VW factory. Translation:
1. Volkswagen—why, 1.) Motorization, 2.) Luxury or transportation, 3.) Quantity and price, 4.) Sensible control of demand
2. Instead of food culture, Volkswagen and remaining car production, No competition, Just: One Volkswagen
3. Cornerstone, Germany's largest car factory, Factory housing community exemplary

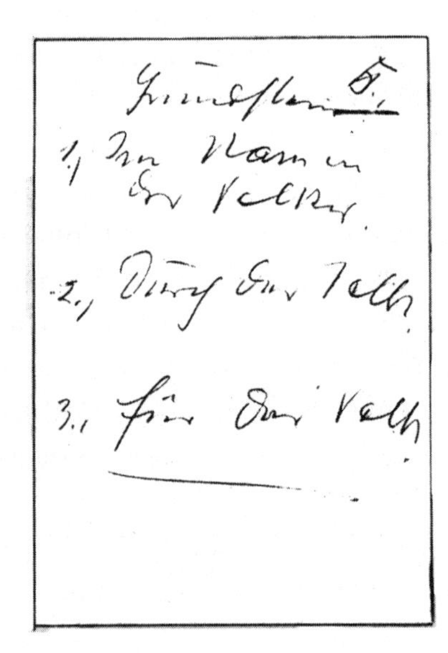

Fourth and fifth note, out of a set of five, created by Hitler in preparation for his speech at the opening of the VW factory. Translation:

4. 3 Men 1.) Werlin, 2.) Loverrenz [Lafferentz misspelled], 3.) Porsche, Dr. Ley—KdF

5. Cornerstone 1.) In the name of the people, 2.) By the people, 3.) For the people

production stage, would be built. In Germany, the car was also known as the KdF-Wagen, since the project was supported by the Nazi organization Kraft durch Freude (KdF), "Strength through Joy," part of the Deutsche Arbeitsfront, the German Labor Front.

On the train to Fallersleben, Hitler had made some notes for his opening speech. On five blank cards he wrote down the most important features of the Volkswagen and the names of the people he wanted to thank in his speech, including Jakob Werlin and Ferdinand Porsche. Hitler extemporized his entire speech to the many dignitaries at the opening ceremony using just the keywords on those five cards. One aspect he emphasized was that there would be "just one Volkswagen" and "no competition."

The first twenty thousand vehicles would start rolling off the production line in September or October 1939. The standard model was to cost RM 990; a large folding roof could be had for an extra sixty. Tens of thousands of German workers faithfully stuck five-Reichsmark savings stamps onto their savings cards so that they could become the proud owners of a real Volkswagen. None of them would ever receive a car.

Patent Battle in Germany

After Ganz sold his most important patents to ERFIAG in 1937, Ferdinand Fuchs took steps to prevent their violation. One of his first cases was against the French manufacturer Citroën, which since 1936 had been equipping its Traction Avant model with an advanced rack-and-pinion steering system patented by Ganz, without paying royalties. A second case was started against Daimler-Benz in Germany over a pat-

DEUTSCHES REICH

AUSGEGEBEN AM
25. FEBRUAR 1937

REICHSPATENTAMT

PATENTSCHRIFT

№ 642 192

KLASSE **63**c GRUPPE 38_{03}

D 63370 II/63c

Tag der Bekanntmachung über die Erteilung des Patents: 11. Februar 1937

Daimler-Benz Akt.-Ges. in Stuttgart-Untertürkheim

Kraftfahrzeug mit vier unabhängig abgefederten Rädern

Patentiert im Deutschen Reiche vom 22. April 1932 ab

Header of DE patent #642192 for independent suspension on a car with a rear mounted engine, filed by Daimler-Benz in Germany, 1932

ent for independent suspension[31] that Daimler-Benz had applied for in 1932 before using it in its Mercedes-Benz models 130H and 170H. When this patent was finally granted in 1937, Ganz submitted an objection, saying that it was based on his design for the Maikäfer and therefore legally his property. He was supported by Fritz Isler, who had taken a test drive in the Maikäfer in July 1937. On looking at the drawings, he concluded that "in terms of patent rights this corresponds in every respect with the rights defined in German patent DRP 642192" and said Ganz could claim it as his intellectual property.[32] ERFIAG started a court

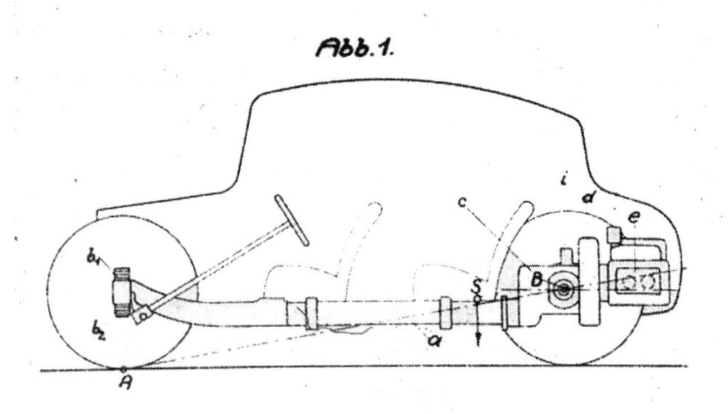

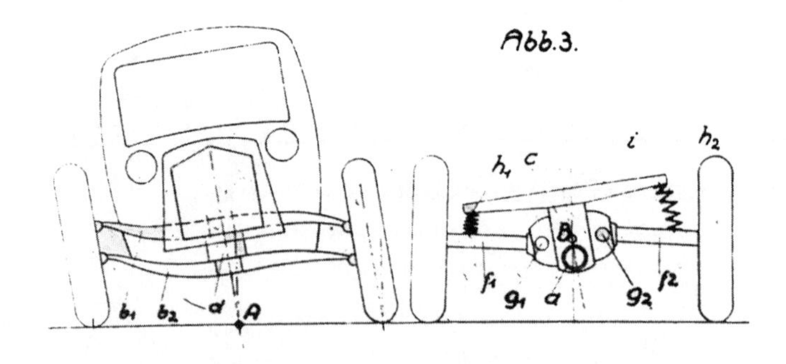

Drawing from DE patent #642192 for independent suspension on a car with a rear mounted engine, filed by Daimler-Benz in Germany, 1932

case against Daimler-Benz but reached a settlement with the company in August 1938. In return for one Swiss franc for each car produced, ERFIAG would be given a license by Daimler-Benz for all countries in which the patent was valid. One manufacturer that would not need to pay for the patent was the German Volkswagen factory, which, by order of Hitler, could make whatever use it pleased of patents held by German car manufacturers.[33]

Another court case was started against Ambi-Budd Presswerke GmbH, a manufacturer of pressed-steel bodywork, which ERFIAG claimed had violated a Ganz copyright on a chassis design in producing various models for BMW and Adler. August 1938, however, saw a major setback to the copyright cases. Josef Ganz was officially stripped of his German citizenship, and all his property in Germany was confiscated.[34] This meant he could no longer make any claims on his German patent rights. Less than a month later, a dramatic verdict was issued by the court in Berlin: ERFIAG was ordered to pay a staggering RM 200,000 to Ambi-Budd for rights to an invention originally patented by Ganz. If it failed to do so, a prison sentence of six months would be imposed.[35] ERFIAG appealed against the decision, but its appeal was rejected.[36] After a second appeal, ERFIAG was forced to pay only the legal costs of the case. To compensate itself for this revised decision, Ambi-Budd laid claim to all of Josef Ganz's German patents.

War in Europe

In 1939, plans were drawn up for serial production, initially of fifty to seventy Swiss Volkswagens, by Rapid Motormäher AG in Zürich. The retail price of the car was fixed at two thousand Swiss francs. With an eye to foreign license agreements, a press release was compiled in German and English with an extensive description of the car, innumerable quotes from the media, and an appendix outlining all of Ganz's patents.[37] The press release stated that the car would be sold for only RM 600, almost four hundred Reichsmark less than the German Volkswagen. In the summer of 1939, license negotiations for the German and Swiss Volkswagens intersected when German negotiators in Turkey were handed a press release describing the competing Swiss Volkswagen developed by Ganz, along with three photographs of the vehicle. They took the material back to Germany, where it was passed to Ferdinand Porsche to study.[38]

ERFIAG made its first license agreement in July 1939, with the First Polish Locomotive Factory, based in Warsaw and Cracow. While it was being negotiated, Paul Ehrhardt was spotted in Switzerland again, probably having come to sabotage the development of the Swiss Volkswagen. He was supported by Margaritta Röhr—the widow of Gustav Röhr, who had died on August 10, 1937—with whom Ehrhardt was in close contact. Ehrhardt visited Switzerland frequently, but because of statements submitted by Josef Ganz in February 1939 he had already been arrested and interrogated once by the Swiss police. Ehrhardt had been released, but he was still a suspect. When he turned up in Switzerland again in early July 1939, he was once more arrested and questioned. Ehrhardt wrote afterwards to the immigration police that the accusations against him originated with "an émigré German Jew, Josef Ganz," the former editor-in-chief of "one of the sensation-seeking 'trade journals' that have unscrupulously fleeced the German car industry." About his own role in the matter, Ehrhardt wrote: "When an end was being put to his practices, I contributed out of a sense of duty." Ehrhardt declared to the immigration authorities that Ganz had told lies about him to the police, as well as reporting him to the Czechoslovak Ministry of War as a spy, and had thereby abused the "generously upheld right to asylum." "The criminal instincts that are innate to Ganz, of which a small selection is presented here, are developing within the safety of his asylum into poisonous outgrowths of which a small selection is presented here."[39]Ehrhardt's efforts were fruitless. In August 1939, the Swiss Volkswagen was demonstrated by Ganz to representatives of the French Ministry of Defense on its testing ground for army tanks in Vincennes. The test drive went so well that within two days a license agreement had been signed with the French auto manufacturer Rosengart for the production of 25,000 cars, for a fee of two hundred French francs for the first five thousand and a hundred francs for each subsequent car.[40]

Hitler's war put a stop to production plans in France and Poland. Having annexed Austria in March 1938 and occupied Czechoslovakia in March 1939, Germany invaded Poland in September 1939, plunging Europe into a state of war. The glistening new Volkswagen factory in Fallersleben immediately switched to the production of military off-road and amphibious vehicles, which Porsche had developed over the past few years based on his Volkswagen design. In April 1940, the occupation of Denmark and Norway followed, and in May the German army invaded France, Belgium, Luxemburg, and the Netherlands. With the occupation of large parts of Europe, Germany had closed the net around neutral Switzerland and was able to exert far greater political pressure on it than before.

Libel from Two Camps

Paul Ehrhardt and Margaritta Röhr found a willing ear for their libel campaign against Josef Ganz in the Swiss police detective Heinrich Schoch. In his letters and reports, Schoch spread exactly the same lies as Ehrhardt had perpetrated years earlier in Germany: the Maikäfer was not a viable concept, Josef Ganz was not a capable engineer, and *Motor-Kritik* was nothing more than a sensationalist rag that Josef Ganz, along with his colleague Frank Arnau, had used to blackmail the German auto industry. It was clear whom Schoch meant in his letters to the Swiss security service when he referred to a "confidant" who had supplied this information. He even tried to use Ganz's relationship with Madeleine Paqué to blacken his name, saying he was living in sin with his Roman Catholic "bride." Because he was now once again deemed a suspect, all mail to and from Josef Ganz went through Schoch's office.[41] It included letters between Ganz and Jean Grünbaum, a Czechoslovak auto engineer of Jewish extraction who had worked for a time on the development of the Swiss Volkswagen. Grünbaum had traveled to Geneva on March 10, 1939, merely to visit the motor show there. Five days later, the German army invaded his country, and Grünbaum remained in Switzerland. He was engaged by ERFIAG to work on the Swiss Volkswagen, but a year later renewal of his work permit was denied, and in September 1940 he was interned in Davesco, a Swiss labor camp. In March 1941, Grünbaum emigrated to South America with a group of engineers and scientists. They invited Josef Ganz to go with them, but he did not want to abandon his work and in any case regarded staying in Europe as a minimal form of resistance to Hitler—unaware that in Switzerland the same fate awaited him as had befallen Jean Grünbaum.

Behind the scenes, the police detective Heinrich Schoch was supported in his campaign of libel against Ganz by one Herr Wälchli of the Ministry of Industry, Handicrafts, and Labor, which together with director Arnold Rütishauser of Rapid wanted to lay claim to Ganz's Volkswagen design. In February 1941, Wälchli wrote to the immigration police that the patent on the Swiss Volkswagen should be transferred to Rapid and that the company had now reached the point where it could build five Swiss pre-production cars. This meant, he said, that "the problem of the "Swiss Volkswagen" has now happily passed into purely Swiss hands." Wälchli also announced that Rütishauser and not Josef Ganz was the true spiritual father of the Volkswagen and that Ganz was not even "capable of bringing a technical design to successful completion."[42] As a direct consequence of Wälchli's letter, Ganz's work permit and residence permit were withdrawn by the Swiss authorities. Less than a month after Wälchli wrote to the police, Ganz received an official letter saying he must leave Switzerland within one month, by April 20 at the latest. With the help of his lawyer Ferdinand Fuchs, he managed to secure permission to stay until May 31, then until July 21, 1941.

Just over a month after the final deadline expired, Ganz received a police order to report to a Swiss labor camp for immigrants, where he would be interned.[43] Like Jean Grünbaum, his sister Margit had been sent to a work camp in Schaffhausen near the German border and interned there after she fled to Switzerland with two suitcases in response to the German occupation of Austria. She had managed to escape and was in hiding with the writer Felix Salten, Austrian by birth and an old Jewish friend of their father Hugo Ganz. In 1936, Hitler had banned Salten's work, which had found international fame with the filming of his book *Bambi* by Walt Disney. Like Margit, Salten had fled to Switzerland after the annexation of Austria in 1938. He had not yet been interned, and Ganz had managed repeatedly to avoid that outcome. Ganz had the good fortune to be working on various inventions in which both the Swiss government and Swiss industry were extremely interested, including a lubrication system for pistons in engines and pumps,[44] a system for sharpening twist drills used in tunneling,[45] a drive mechanism for a bicycle that was less tiring since the pedals moved in an ellipse,[46] a suspension system for the bicycle,[47] and special cutters called Cupex to snip ration coupons out of the middle of large-sized ration books, along with the Cupofix, with which to organize them. During the war, with the help of his nephew Felix Ganz, he put the cutters into production and spent whole evenings at home assembling them along with Madeleine Paqué, Ute Klüpfel, and little Dieter. As long as he was developing highly original devices of this kind, the Swiss government would not deport or intern him for fear of losing valuable inventions.

One remarkable "coincidence" was that the internment order reached Josef Ganz only three days after the final verdict in the court case against Tatra. After legal proceedings lasting over seven years, the accusations by Tatra turned out to be so weak that the court in Berlin ultimately found in favor of the émigré Jew Josef Ganz. Tatra was ordered to pay him more than RM 4,000.[48] The fine was judicially valid, and steps would be taken if necessary to make Tatra pay. All this time, Wilhelm Gutbrod had avoided having any contact with Josef Ganz in view of the need to preserve "the good name" of his company and "on grounds of propaganda." Since 1934, Ganz had missed out on royalty payments for around 1,200 cars a year.[49] In his case against Tatra, Ganz had been represented by the famous lawyer Dr. Julius Israel Fliess of Berlin. As a result of his services in the First World War, the much decorated Fliess found himself in the unique position of being allowed to continue to practice in Germany as a Jewish lawyer, although in 1938 he had been obliged to add "Israel" to his name. Tatra vs. Ganz would be one of his last cases in Germany. In September 1942, he fled to Switzerland with his family, each of them disguised as an agent of German military intelligence, the Abwehr, as part of Operation U-7, organized by German resistance fighters Hans van Dohnanyi and Friedrich Arnold. He had narrowly escaped death in a concentration camp.

Enjoying the respect of Hermann Goering, the aviation pioneer and Tropfenwagen-designer Edmund Rumpler had also been able to remain in Berlin throughout the 1930s, going against his wife's dramatic pleadings to emigrate and declining "incredible offers" from the United States. By the time Rumpler had realized what Hitler was, it was too late. He was not allowed to leave Germany because in the fraternity of German engineers he "knew too much" about what they were after. Although unlike Fliess he was not allowed to work, Rumpler had kept working on some engineering projects in secret, including a design for a small Volkswagen. In the case of Rumpler, it was his death on September 7, 1940, that saved him from a far more tragic and gruesome end. After Rumpler died, the Nazis threw out his entire archive, which included every drawing and blueprint he ever made, out of the window of his home in Berlin into the courtyard below to be picked up as trash and destroyed. All that remained was the first-built Tropfenwagen that Rumpler had donated to the Deutsches Museum in Munich years before his death, in 1925.[50]

Among the many millions of European Jews who did not escape were several of Josef Ganz's German relatives. His cousin Elisabeth Ganz was killed at Auschwitz extermination camp in Poland in October 1942, and his aunt Regina Ida Ganz died at Theresienstadt in Czechoslovakia in March 1943, on her way to Auschwitz. Jakob Feldhammer, Margit's ex-husband who had been of so much help to Ganz in 1933 after his arrest by the Gestapo, had been in an Italian concentration camp since 1941. In May 1944, he was transported to Auschwitz and gassed there within a week. Although throughout the war Josef Ganz was relatively safe in Switzerland, he suffered terribly from the perpetual uncertainty about his temporary residence permit as well as the intrigues entered into by the Germans and the Swiss, which might at any moment lead to his imprisonment in a labor camp or deportation to Germany. Ganz regularly suffered severe migraine attacks. They had started after his car accident in 1930 but were made worse by the problems he faced daily. Often he sat at home glued to the radio to hear news from Britain about how the war was going. The most difficult periods in those war years were the cold winter months. Josef Ganz, Madeleine Paqué, Ute Klüpfel, and Ute's young son Dieter would live mainly in the passageway, the only space in the apartment where they could keep reasonably warm with the minimal amount of fuel they were able to feed into their little stove.

Switzerland for the Swiss

Ferdinand Fuchs of ERFIAG, facing problems with Rapid Motormäher AG and the loss of all revenue from license agreements in France and Poland as a result of German occupation, tried to secure new license agreements in Switzerland for production of the Swiss Volkswagen. In 1942, he loaned one of the prototypes to engineers from the Spanish-Swiss car firm Hispano-Suiza in Geneva, who spent two weeks studying and extensively testing the car. They were extremely interested in Ganz's design, which they judged to be "ingenious, simple, and sound." Hispano-Suiza even made calculations as to what a complete factory, including steel presses, would cost, should they choose to put the Ganz Volkswagen into production.[51] The severe shortage of raw materials and funding in wartime, however, made it impossible to get the project off the ground.

Although he carried out several more test drives with the Swiss Volkswagen, it was early 1944 before Ganz seriously picked up his design work again, when the downfall of the Third Reich finally began to seem a realistic prospect. Ganz laid out his old designs for a streamlined Volkswagen on his drawing table and started to modernize them for Hispano-Suiza. He was hoping that after the war the closed Volkswagen with four seats could go into production in Spain. The design process was still in full flow in January when the doorbell rang. At the door were two detectives, Otto Dätwyler and Otto Gloor of the Zürich police. Police Commissioner Frei, who had dealt with the first investigation into

Ganz nine years before, had ordered them to look into his latest application for the postponement of his expulsion deadline. The two detectives searched the house and tore the drawings off the table as "evidence." In their incriminating report they called Josef Ganz an "undesirable alien" who was good at "approaching interested parties and wheedling large sums of money out of them." Despite "debts of around 28,000 Swiss francs," Ganz was said to live "in grand style," driving around in his Volkswagen on gas rations intended for test drives even though he had a season ticket for the tram, "a mode of transportation with which this 'parasite' is not content." They said he did not possess any "outstanding qualities as a technician." The old drawing on his table was "proof" that he had not developed anything new in ten years.[52] That same day, in the mail, Madeleine received an invitation to join the Nazi Party—a clear sign that Josef Ganz was being sabotaged by people in Germany.

In late March 1944, Ganz was fined 350 Swiss francs for ignoring a work ban imposed by the immigration police. It was the extremely insulting police report by Dätwyler and Gloor, however, that proved the final straw. With the help of Ferdinand Fuchs, he appealed against the fine and decided to pass the entire case to the press via a journalist friend, Dr. Fritz Heberlein. On June 29, 1944, articles appeared in the Swiss newspapers *National Zeitung*[53] and *Tages-Anzeiger für Stadt und Kanton Zürich*[54] disclosing the libelous police report about the "parasite" Josef Ganz. The next day an unsigned postcard landed on Ganz's doormat on which was written: "Jews out!! Switzerland for the Swiss! First warning!!" Now that all the difficulties Ganz had faced over the years had been made public in articles in the *National-Zeitung* and the *Tages-Anzeiger*, there was a whirlwind of reports in the rest of the Swiss press. In late January and early February 1945, *Volksrecht* published a four-part article headed "Immigration Police in the Service of the Gestapo?"[55] It openly discussed the roles of "German spy" Paul Ehrhardt and of Wälchli, the man who in 1941 had tried to ensure Ganz's Volkswagen designs were in "purely Swiss hands." As the Second World War finally came to an end after five long years and the loss of millions of lives, a fierce battle broke out in the Swiss courts between Josef Ganz and his opponents.

CHAPTER SIXTEEN

SANCTUARY IN AUSTRALIA

On the morning of March 4, 1945, Zürich was rocked by a series of huge explosions. In poor weather, American bombers on their way to Germany had mistakenly dropped twelve tons of incendiary bombs and 12.5 tons of high explosives on the city. Although most of the ordinance went off harmlessly in an open field, several people were killed, and a number of houses in a street called In der Hub were destroyed, including the one in which Ganz had lived for a while in the mid-1930s.[1] It was not his first narrow escape, and Josef enjoyed telling the story to his family and friends. This accidental bombardment of "neutral" Switzerland took place in the closing weeks of the war. Less than two months later, Russian troops advanced through the ruins of central Berlin, where Adolf Hitler, deep in his underground bunker complex, committed suicide on April 30, 1945. Just over a week after that, on May 7, 1945, Germany surrendered. The war in Europe was over.

In the following few weeks, fleeing Nazis were hunted down and arrested all across Germany and beyond. Paul Ehrhardt turned up in Denmark, where he pretended to be a member of the Red Cross who had saved a Jewish family and claimed he had been involved in the assassination attempt against Hitler by a group of German officers led by Colonel

Claus von Stauffenberg in July 1944. Everything he said was refuted in an investigation carried out by Alfred Bloch of the Economic Intelligence Division, an arm of the American occupying force in Berlin. Not only did Bloch find important witnesses willing to testify against Ehrhardt—chief among them a German colonel who soon afterwards committed suicide "in mysterious circumstances"—he also had at his disposal an extensive personal dossier of Ehrhardt's, found by the US Army in July 1945 in the ruins of the Chancellery among papers belonging to Minister Otto Meissner. Bloch believed Ehrhardt was a highly placed SS general, decorated by Hitler with the Nazi Party's Gold Medal for "special services" rendered at the Mercedes-Benz factory, "in the construction of a car (presumably the *menschensvergasungsauto*)"—in other words, for designing a car used to gas people.

Bloch, a close acquaintance of Josef Ganz, passed the dossier to the war crimes court in Nuremberg, headed by Robert Jackson.[2] But Ehrhardt was never tried for his crimes. He had set himself up as an informant to the US Army and came under its protection through the good offices of Frank Röhr, son of the late Gustav Röhr, who had joined the American armed forces and worked for the American car manufacturer General Motors. Ehrhardt was given a place of safety in Frankfurt, heavily protected by US forces, and the dossier on his case "disappeared." Switzerland was the only country in which details of his crimes were openly published, in newspaper articles about Josef Ganz.

Jakob Werlin was arrested by the American armed forces in 1945 and imprisoned for having held various high positions in Nazi Germany, including those of General Inspector of Motor Transport, Member of the Management Board of Daimler-Benz, and Member of the Board of Directors of Volkswagen. Werlin was seen as "one of the main promoters of the Volkswagen saving scheme in which a great number of people made regular payments to the Labor Front on promise of future delivery of the car." However, "the Volkswagen was never built and the contributions of the people helped instead to finance the construction of military vehicles at the Volkswagen plant in Fallersleben."[3] As "one of Hitler's most intimate friends," Jakob Werlin was even questioned about whether Adolf Hitler was alive or dead.[4] Josef Ganz's close friend Georg Ising testified against Jakob Werlin as a "particular usufructuary of the Hitler Regime." Ising explained in great detail to the Military Government in Munich how Jakob Werlin had persecuted Ganz so as to "make it impossible for him to build and utilize the true Volkswagen." Ising wanted to "rehabiliate the life's work of Josef Ganz" and ensure he was given some form of compensation from the personal assets of Jakob Werlin for the injustice he had suffered.[5] But to no avail. Jakob Werlin was released from prison in 1948 as nothing worse than an opportunist under the Nazi Regime and resumed his original work as a car salesman for Mercedes-Benz in the German towns of Rosenheim and Traunstein.

Immediately after the war, the media reported on the terrible crimes against humanity committed by the Germans in the concentration camps. A good friend sent Josef Ganz a series of blurred photographs taken secretly in one of the camps during the war. It showed Jewish prisoners being tortured and hanged. Ganz was horrified to learn that the photographs came from the Swiss ambassador to Germany, who had been in possession of them since 1942. He was so terrified of the Nazis that he had never done anything with them. Instead, concerned about his own safety, the ambassador had handed them over to Ganz's friend "for safe keeping."[6]

Decline and Rise of the Volkswagen

Meanwhile, the Swiss Volkswagen project had more or less run into the sand. Keen to be able to use the design, Ganz's intellectual property, without paying for a license, the director of Rapid, Arnold Rütishauser, had all the drawings made and signed by Ganz copied onto the official drawing paper of Rapid Motormäher AG to disguise their origins. Technical assistant Leo Widmer of Rapid watched all kinds of changes being made to them. Rütishauser—who "is definitely not an auto technician"—even ordered many "carefully considered details" to be altered.[7] Rütishauser then said he had been forced to redevelop Ganz's designs from scratch because they were defective.[8] The pre-production model was completed in the summer of 1945, but in 1946 and 1947 only a few dozen cars were built. Meanwhile, Ganz continued to carry out test drives in the Silverfish, the prototype with aluminum bodywork that he had perfected himself. He went looking for increasingly extreme terrain to test it on, such as woodland with tree roots snaking up over the ground.

By contrast, the German Volkswagen factory, parts of which had been reduced to rubble by Allied bombing, had recommenced production in 1946 under direct British control, making a virtually unaltered pre-war Volkswagen. While cars began rolling off the assembly line in their thousands, former colleagues of Ganz in Germany tried to restore his reputation as the pioneer of the Volkswagen. Georg Ising, who had worked as editor-in-chief of *Motor-Kritik* until the end of the war, relaunched the magazine in 1947 under the new title *Motor Rundschau.* In March, in the fifth issue of the year, he wrote that as part of efforts "to prove that only 'Aryan' people were capable of creative deeds, not only was history falsified on a grand scale in the Third Reich, countless falsifications also took place in the technical field." Ising said that "for example, a great deal was written on the subjects of 'streamlining' and 'Volkswagen,' which are now very much in the news again, while the names of those who must be deemed the true originators and pacesetters were continually and consistently suppressed." He gave their names: "in the field of streamlining research Chief Engineer Paul Jaray and in the field of the Volkswagen Qualified Engineer Josef Ganz."[9] Porsche's "German Volkswagen," which Hitler had promised would be marketed for only RM 1,000, now turned out to cost five times as much to produce. "A Volkswagen characterized by the words 'as comfortable as a car and as cheap as a motorcycle' is not what we have," Ising wrote, "although such a car is what we urgently need, both now and in the time to come." That design—Josef Ganz's Swiss Volkswagen, introduced to readers of *Motor Rundschau* by an extremely enthusiastic Georg Ising—would never go into production.[10] After test runs with the pre-production series, the Swiss version was found to have too many teething problems, caused by the changes made to Ganz's original design by Rütishauser of Rapid. During the war, the former co-editor of *Motor-Kritik* Frank Martin had been editor-in-chief of *Das Auto*. He visited Ganz in Switzerland in 1948 and took the Silverfish for an extensive test drive. Of the German Volkswagen he wrote: "There can be no doubt that the People's Car unjustifiably bears that name, given that so few are as yet able to afford one." He concluded that Ganz's design was for precisely the kind of cheap all-purpose vehicle so many people needed in these difficult postwar years.[11]

Heading for the article by Georg Ising in *Motor Rundschau*, 1947

Meanwhile, Josef Ganz and Ferdinand Fuchs were trying to arrange new licensing deals for the production of the Swiss Volkswagen, and on June 24 and 25, 1948, they demonstrated the car to representatives of the Italian company Societá Commerciale Aeronautica MACCHI in Milan. The manufacturer expressed a desire to produce a pre-production series of a hundred followed by a production series of five thousand cars.[12] It never did so. Besides Italy, Ganz and Fuchs explored contacts in France, where the firm Automobiles MA Julien of Toulouse showed interest. It worked on developing several prototypes, and Ganz traveled regularly to France to see how the project was progressing. In the same year, he developed an innovative suspension system for skis. It was used on the bobsled with which the Swiss team won gold and silver medals at the 1948 Winter Olympics in St. Moritz.[13]

Complex Legal Battle

Josef Ganz's attempts to resume his work were overshadowed by increasingly complex legal difficulties as a result of the police report written by Otto Dätwyler and Otto Gloor in 1944. A stream of articles on the subject in the Swiss newspapers in 1947 had led the two detectives to begin a case against Ganz and his lawyer Ferdinand Fuchs for defamation. It would drag on for years, receiving widespread coverage in the Swiss press and leading to a series of related court cases. The first follow-up move came when Dätwyler and Gloor declared in court that they had based their investigation mainly on an older report by the police detective Heinrich Schoch, who in turn said that his report was based on statements by Margaritta Röhr, the widow of Hans Gustav Röhr. In January 1948, the Swiss newspaper *Freies Volk* made much of the involvement of Margaritta Röhr—the alleged mistress of Paul Ehrhardt and an active participant along with him in the libel campaign against Ganz. The paper called her deceased husband a "swindler."[14]

Margaritta Röhr was not the type to take such remarks lying down, and along with her son Frank Röhr she too started a case against Ganz for defamation. Among her supporters were Paul Ehrhardt and the German journalist August Christ. Josef Ganz and Ferdinand Fuchs then began a case against Jakob Krieg, Ganz's upstairs neighbor, who had once worked on the Swiss Volkswagen and had been used by Ehrhardt as an outlet for lies about Ganz. They accused Krieg of giving false statements during the case they were fighting against Dätwyler and Gloor.[15] This new case too would drag on for several years, and witness after witness was called to testify for and against Ganz and Krieg, roughly a hundred of them in all. Among those called by Krieg were Paul Ehrhardt and Margaritta Röhr.[16] After fifteen years of harassment by him, Josef Ganz now stood eye to eye with his archenemy. Ehrhardt made a number of false statements to the court, repeating the old lies that Ganz had blackmailed the German auto industry with his "sensationalist rag" and had based his Volkswagen on ideas first set down by Tatra and by Ehrhardt himself. He was unmasked by the court as a liar.

Gradually, more and more people became caught up in the legal battle. By the late 1940s, innumerable cases were being heard, pitting Josef Ganz against opponents who included several people in Swiss government institutions. Most of the allegations concerned defamation, giving false statements in court, or withholding evidence from previous court cases. A reporter at the Swiss paper *Die Tat*, who like many journalists supported Ganz, remarked that the legal tangle surrounding Josef Ganz had become so vast and complex that it "threatened to paralyze the entire judiciary."[17] On top of all this, in late 1948 Ganz had submitted claims to the German courts on a number of grounds: loss of income as a result of the declaration that four of his German patents were null and void; the racially motivated termination of his contracts with Daimler-Benz, BMW, and Standard Fahrzeugfabrik under pressure from the Gestapo; his right to compensation for the chassis built by Ambi-Budd that relied on a design he had patented; and his entitlement to Daimler-Benz's patent DRP 642192, which he said was based on his own inventions. He demanded royalties for the German Volkswagen and the Mercedes-Benz models 130H and 170H, which made use of that patent.[18]

Meeting with Porsche

This enormously complex mass of court cases put Josef Ganz under so much pressure that it cost him his relationship with Madeleine Paqué, his companion of twenty years through thick and thin. Amid the problems of the late 1940s, Ganz got to know a nurse in Zürich called Heta Jacobson, and to Madeleine's great sorrow he began a relationship with her. They would not be able to spend much time together in those turbulent years, however. Ganz was tremendously busy dealing with the demands made on him by the legal rigmarole and his regular trips to France, where he had become intensively involved with the development of a series of prototypes by Automobiles MA Julien, based on a concept very similar to the Swiss Volkswagen. Two versions were developed, the MM5 and the MM7, the latter having a more powerful engine and slightly different bodywork. In November 1950, they went into pre-serial production, and eighty cars were built.[19] Heta Jacobson left Switzerland for good that year and followed her sister Erna by emigrating to Australia.

The avalanche of court action came to an end in late 1950 without any satisfying results. Several cases were decided in Ganz's favor, but in one, brought by a Swiss public prosecutor, he was sentenced to three weeks in jail for defamation and giving false statements, since he was unable to prove his accusations of a Nazi plot. In the Swiss press, which was sympathetic to Ganz, the jail sentence prompted serious doubts about his conviction. The sentence was reduced on appeal, but it was probably because of pressure from the prosecutor and other highly placed people in Switzerland that Ganz received the cruelest blow of all. He learned in October 1950 that his residence permit, which had been extended at his request every year since 1935, would not be renewed.[20] On October 9, 1950, he was arrested by the police and ordered to leave Switzerland that same evening, since he no longer had a valid permit to remain. The reason given for this rapid sequence of events was that Ganz was "a Communist secret agent,"[21] but soon people were saying he had actually been expelled from the country as a "troublemaker." A journalist for the Swiss paper *Dimokratikos* went so far as to declare that Ganz was the victim of a top-secret "National Socialist spy network" in Switzerland that "is without any doubt in contact with highly placed industrialists and financiers in West Germany."[22] Ganz hurriedly took his leave of his old love Madeleine Paqué and drove flat out to Paris via Geneva in his gleaming aluminum Silverfish. In Geneva he met with Ferdinand Porsche, who struck Ganz as looking like a man who had "been through a great deal." After the war, Porsche had spent twenty-two months as a prisoner in France. The two men talked about the latest developments in which they were involved and parted on good terms.[23] It was their final meeting. Porsche died on January 30, 1951, at the age of seventy-six.

Ganz traveled on to Paris, where he would live for several months at the simple Hotel Max on the Avenue de Malakoff in the sixteenth arrondissement. His problems in Switzerland followed him there, and he was arrested and interrogated in Paris but released after managing to refute allegations that he was a "Communist secret agent."[24] His health had suffered so badly in recent years that he had a heart attack in Paris and was admitted to a hospital there for several weeks. When he left the hospital, his beloved sister Margit came from Switzerland to visit him. She was shocked by the straitened circumstances in which the once-wealthy Josef was living.[25] He would never see her again.

In his simple hotel room in Paris, Ganz contemplated his impossible position. He could not return to Switzerland, where for years he had tried to build a life for himself, and he felt threatened by former Nazis like Paul Ehrhardt who were still free men. He could see no future for himself in "liberated" Europe, and as soon as he had recovered he went to the Australian Embassy in Paris to find out how he could emigrate to the other side of the world to be with his new girlfriend Heta Jacobson. On the Boulevard du Palais he stepped into a Photomaton, a booth with a machine that promised six passport photos within eight minutes. He made two sets, one full-face and one in profile. He looked distinguished with his gray beard, but his expression was weary. In just a few years he had become an old man. A short while later, he took a train to Marseille, and that summer he boarded the ocean steamer the *Chang Chow*, bound for Sydney, Australia.

New Life in Australia

After two months at sea, the ship reached Sydney harbor on September 17, 1951. Fifty-three-year-old Josef Ganz was one of hundreds of passengers. After almost twenty years of persecution in Europe, he was seeking sanctuary in Australia. From Sydney, he traveled on to Melbourne in the state of Victoria, where he moved in with his girlfriend Heta Jacobson and her sister Erna, who lived in St. Kilda, a small coastal suburb of Melbourne. Ganz found work in Melbourne as a technical designer at the rubber factory Rubbertex Pty Ltd. One of his projects there involved the development of safety features for the hazardous machine tools used in the factory. His maverick way of working made Ganz an oddity at Rubbertex. To the annoyance of his boss, he sometimes spent days or weeks walking around the factory talking to the workers. Although it seemed as if he could not possibly be doing anything concrete, Ganz would then spend a mere few days making fully detailed technical drawings for astonishingly simple but ingenious factory equipment.[26]

Although he was able to create a number of important safety devices for Rubbertex, his heart still lay with auto engineering, and he went in search of work in the Australian car industry, such as it was. He found a job with the Standard Motor Company (Australia) Pty Ltd, which was mainly engaged in assembling cars made by its British sister company Standard. Occasionally, models were adapted for the Australian market; Ganz worked on the development of a pickup, convertible version of the Standard Ten. He fairly soon left the company, however, and in 1954 joined the Engineering Department at Holden in Melbourne, the Australian branch of America's General Motors (GM). With his impressive technical qualifications and extraordinary background, Ganz was taken on as head of the Special Projects Department and worked at designing suspension systems and developing new prototypes. He was at the same level in the hierarchy as Assistant Chief Engineer George Quarry, reporting only to Chief Engineer G. R. Lewis, head of the Engineering Department,[27] which was fairly small and based in a factory building that had been converted into office space. Ganz was given a small office in a corridor next to the company library and the workshop entrance. The room was just big enough for a small desk and a drawing table. His work at GM Holden was of no great consequence compared to the innovative and trailblazing developments for which he had been responsible in Germany before the war. The factory did not even develop its own cars but produced adapted versions of models made by GM in the United States and Opel in Germany.

Out of loyalty to his employer, Ganz bought a secondhand FJ Holden, but he was not particularly impressed by the car. In fact, he enjoyed driving it on unpaved roads to demonstrate to passengers how bad its suspension and road holding were, but at least it got him around. In his FJ Holden he visited various motoring events, including the popular REDeX Reliability Trial in 1955, where dozens of cars from firms such as Holden, Peugeot, and Volkswagen competed in exhaustive tests. At his age, Ganz no longer took part in races, but he still had a fairly reckless approach when it came to auto engineering. During a test drive in a GM Holden prototype in 1957, he asked the driver to keep the car at a constant high speed while he looked under the dashboard to try to trace the source of a rattling noise. The driver did as he was told but drove straight over a deep pothole, and Ganz hit his head hard on the underside of the steel dashboard. It was almost an exact repeat of his accident with the Röhr Cabrio-Limousine in Germany almost thirty years earlier. He had a large head wound and severe concussion, which forced him to rest for many weeks, first in a hospital and later at home. From his sickbed, he corresponded with family, friends, and acquaintances in Europe, including Madeleine Paqué and Georg Ising. In the summer of 1958, he received from Ising the sad news that his faithful old friend and colleague Walter Ostwald had died at the age of seventy-two.

Among the books he read in bed was *Beyond Expectation: The Volkswagen Story* by the journalist and engineer Kurt Bernard Hopfinger, published in the 1950s based on conversations with Ferdinand Porsche and Volkswagen's director Heinrich Nordhoff. The book told the story of the People's Car, but in a one-sided, incomplete manner that romanticized Porsche's "dream of a Volkswagen." Ganz was not named at all. The rapidly increasing popularity of the VW Beetle—in 1955, Volkswagen produced its millionth Bug—ensured that Hopfinger's book sold well worldwide, becoming the standard work on the history of Germany's "economic miracle." As a one-time reader of *Motor-Kritik* and a close acquaintance of Josef Ganz, Heinrich Nordhoff knew how great Ganz's influence on the pre-war German car industry had been and wanted to give him some form of credit and recognition. When on November 13, 1958, Nordhoff accepted on his own behalf and that of the late Ferdinand Porsche the prestigious Elmer A. Sperry Award for Advancing the Art of Transportation, he gave a long speech in which he described the developments in the German car industry that had eventually led to the Volkswagen, designed by Ferdinand Porsche. Nordhoff mentioned no one other than Ferdinand Porsche by name, with one exception, Josef Ganz. He said:

> Many young engineers, particularly in Germany, took up the challenge, and they were spurred on by Josef Ganz, the editor of a magazine called *Motor-Kritik*; he attacked the old and well-established auto companies with biting irony, and led a crusade of new thought with the ardent conviction of a missionary.[28]

Ganz tried to continue fighting his court cases against the German government and various German companies from Australia, but he was largely unsuccessful. The one positive result was that in 1959 his German lawyer Theodor Hartherz came to an agreement with the German state of Hesse whereby Ganz would be paid 30,000 Deutschmark for loss of income.[29] It was a reasonable sum of money, but it bore no relation to the high salary he had been earning before 1933 and ceased to receive because of Nazi machinations. Soon after that decision, Josef Ganz and Heta Jacobson moved into a small apartment in a new building called Edgewater Towers, right next to the beach in St. Kilda. The modernist structure, with its stunning view of the bay, was controversial as one of the first examples of high-rise building in the little coastal suburb.

Tatra Versus Volkswagen

Josef Ganz had heard nothing from his old acquaintance Heinrich Nordhoff for several years and was therefore pleasantly surprised when a letter from the Volkswagen director reached him in early 1961. By "lucky chance" Nordhoff had obtained Ganz's address in Australia from mutual friends and was writing to ask whether he was interested in "helping us out a bit with our work at the Volkswagen factory." He wrote that he felt it to be "most definitely a deficiency" that although Volkswagen kept careful track of specialist journals both German and foreign, "there is absolutely no critical assessment of these publications at all." Nordhoff was himself a former reader of *Motor-Kritik*, and he asked Ganz to let him know what he thought of the idea of making himself useful at Volkswagen with his "great experience and capacity to gain a picture of the entire field of activity."[30] Josef Ganz wrote back that he had been "extremely happy" to receive Nordhoff's letter and that he was "very interested in a collaboration," whether it be "in the manner in which you propose or of a more extensive kind in relation to technical developments." Ganz, at sixty-three, would have liked nothing better than to carry out "a few more years of productive work" at Volkswagen, the company "that helped to open up the Maikäfer idea to the world." He asked Nordhoff to let him know what he "could expect over there."[31]

Under Heinrich Nordhoff's leadership, Volkswagen had grown at an unprecedented rate since 1948. By 1961, well over four million VW Beetles had been produced, and the five millionth was expected to roll off the production line before the end of the year. The Beetle was an enormous success, but with a complicated history. Anyone who could prove that improper use had been made of his designs or patents would be able to put in a claim for millions—which was exactly what Paul Ehrhardt intended to do on behalf of Tatra. Ganz knew that Tatra had started legal proceedings against Volkswagen for violating no fewer than ten of its pre-war patents. It was exactly the same game as Ehrhardt and Tatra had played against him thirty years earlier. Ganz was building up a good relationship with Heinrich Nordhoff, and he now sent him a one-sentence letter reading: "With the Tatra case you definitely need a pilot on board who is familiar with all the reefs and currents."[32] Nordhoff seized Ganz's offer with both hands. There was no one at Volkswagen with such expert knowledge of pre-war auto engineering as Josef Ganz. He would be able to refute Ehrhardt's claims with a meticulously documented analysis of the patents. Ganz sent the relevant material to the Volkswagen patent department for use in the case against Tatra, and meanwhile consultation continued between him and Nordhoff about his forthcoming appointment at Volkswagen. Nordhoff requested, among other things, an indication of what kind of salary he wanted and what sort of accommodation he would like in Germany.[33] But everything was put on hold when Ganz was admitted to hospital in August 1961 with heart problems.[34]

Rumors at GM Holden

In October 1961, the case against Tatra took a turn that was favorable to Volkswagen, and Ganz congratulated Nordhoff by letter. He wrote that their correspondence and the help he could offer Volkswagen against Tatra "brought light in a depressing situation" and expressed the hope that he and Nordhoff would soon be able to celebrate having "slain the dragon."[35] The accusation of violating ten of Tatra's patents had been reduced to a single claim with regard to a patent held by Tatra for an engine mounting.[36] Ganz believed the claim to be unfounded. He said that the design laid down in the Tatra patent was based on a concept described in *Motor-Kritik* in October 1932 by the German engineer Anton Steidle, two and a half months before the Tatra patent application. Volkswagen's success in the case against Tatra came as a direct result of help from Ganz, the star witness who saved Volkswagen from having to pay millions of Deutschmark in compensation. The company was celebrating its twenty-fifth anniversary in October 1961, a milestone marked by the passing of a quarter of a century since the first three Volkswagen prototypes left Ferdinand Porsche's garage in Stuttgart in October 1936. Articles were now appearing in the German press in which the famous German journalist Dieter Grossherr took up the cause of Josef Ganz and Adolf Rosenberger (a Jewish driver and Porsche's financial backer), both of whom had been omitted from histories of the Volkswagen. Grossherr devoted a large proportion of his articles in the newspapers *Frankfurter Rundschau*[37] and *Suddeutsche Zeitung*[38] to the two men's contributions. He also wrote a long series of articles about the history of the Volkswagen for the German "magazine for the Volkswagen driver" *Gute Fahrt*, pointing out: "The man who from 1923 made the publicizing of the Volkswagen idea in Germany his life's work is now almost completely forgotten: Dipl.-Ing. Josef Ganz of Darmstadt."[39]

Ganz was not entirely forgotten, however. Paul Ehrhardt knew only too well what Ganz's contribution had been and that a witness statement from him might scupper the Tatra vs. Volkswagen court case. With the help of Frank Röhr, the son of Gustav and Margaritta Röhr, who worked for General Motors, Ehrhardt spread rumors at GM Holden that in Josef Ganz the company was employing a crook.[40] He engaged the help of a technician at Holden who wanted promotion.[41] Ganz suffered greatly from the mounting distrust at the company. Some of his colleagues were treating him, he felt, like "a murderer out on bail."[42] These latest developments turned him into a nervous wreck. He lay awake night after night.

On December 20, fate struck again; he had another heart attack. He spent Christmas and New Year's weak but slowly recovering in Prince Henry's Hospital in St. Kilda, where he was treated by Dr. Edwin W. Knight. Along with psychiatrists and heart specialists, Knight concluded that both Ganz's "schizophrenic and manic-depressive" emotional state and his heart attacks were a direct consequence of his persecution in Germany and Switzerland.[43] The year gone by had been difficult, but at least the smear campaign and legal action against Ganz had finally come to an end after thirty years. His archenemy Paul Ehrhardt died in 1961.

Ganz was not discharged from the hospital until February 1962. Although the government had granted his request for Australian nationality, he had been unable to attend the naturalization ceremony. It was held later, privately, at his home. He was still very weak and often lay resting in bed or in a chair, with a fan to cool him. It was summer in Australia and sweltering hot. As he grew stronger, he began to occupy himself during his slow convalescence by making a start on his autobiography,[44] using a small traveling typewriter he could balance in his lap, and by resuming his correspondence with family and friends in Germany and Switzerland. To his old acquaintance Dipl.-Ing. Max J. B. Rauck, custodian of the Deutsches Museum in Munich, he sent several letters, enclosing carbon copies of passages from his autobiography. He also wrote to a number of individuals at GM Holden. "It would be the peak of craziness and the biggest waste of energy if we two had to fight each other," Ganz told a member of the board at GM Holden, H. W. Gage, in March 1962. He asked Gage to "please put away all prejudice" and pass on his evidence to someone who could read German. Gage would then, as "the minimum," avoid becoming "involved in the protection of criminals." Ganz promised him "an insight in a crime story even Hitchcock could not invent."[45] He even proposed calling on the Australian Security Intelligence Organisation (ASIO) to carry out an independent investigation into his past.[46] All the evidence needed to support his incredible story had been sent by ship from Zürich to Australia over the past few years by Madeleine Paqué. The pair were still in close contact and sent each other long and emotional letters. The modest apartment in St. Kilda where

Ganz lived with Heta Jacobson was piled high with thousands of documents, magazines, and books; visitors described the place as a library. Since he found it hard to relax mentally, he did not completely recover, and in June 1962 he had another heart attack that brought another stay in the hospital. His recovery, such as it was, took months. Ganz was too weak to return to his work at GM Holden and scotch the rumors about him once and for all.

Volkswagen Pension

A ray of light in those dark days was the birth of Ganz's great-niece Maja von Tolnai. His sister Margit had been updating him for months on developments and "advertised her," Josef said, "almost like a car company before launching a new model onto the market." Two months after her birth, Ganz wrote an emotional letter to his new little great-niece. She had been born on a day he regarded as symbolic, July 20, the date of the assassination attempt on Hitler by Claus Schenk von Stauffenberg in 1944. Josef described himself as an "engineer, just like your dad, although one who has ended up on the other side of the world because he interfered with the wheels of technical history, which at the time stirred a lot of ill feeling." It had been in Hitler's day, "but no one knows what will be written about that in the history books by the time you go to school." Josef expressed the hope that Maja would experience a time "in which it is truly worth living, ruled not by violence but by the law." He would like nothing better than to hold Maja in his arms and to get to know her mother. "I have no illusions, though. Such wishes are almost impossible to fulfill."[47]

Since his last heart attack, Ganz had been practically an invalid, living on a small pension from GM Holden. At Volkswagen, serious plans were afoot in the spring of 1964 to give him a supplementary pension for having "devoted himself at a very early stage, since 1928 or perhaps even earlier, with his entire person and despite many difficulties, to the creation of a 'German Volkswagen.'"[48] The plan was supported by Heinrich Nordhoff and several of his staff, but first Ganz was to receive an initial sum of a thousand pounds in return for his help in the legal battle with Tatra.[49] Partly because of the slow process of international money transfer, the payment took months to reach him and in that period fate struck again. Ganz had another heart attack and was once more admitted to the hospital. From there he wrote a letter to Heinrich Nordhoff, with just one short sentence: "Please help me, I'm in terrible straits."[50] In reply, Nordhoff wrote that although he had felt "somewhat at a loss" when faced by this cry for help, the bank transfer of a thousand pounds was on its way, and he would be happy to investigate "every possibility" of helping him further.[51] Josef Ganz wrote back that the payment of a thousand pounds was a "godsend" that would enable him to leave hospital and "fulfill my most urgent obligations." He was at his wits' end. Since his last heart attack he had been "more dead than alive," so "there can be no question of any kind of work now." Ganz ended his letter with the words: "I am left, as it were, facing nothing."[52] To support Ganz further, Nordhoff sent him a second money transfer later that month worth five hundred pounds.[53]

Despite their good relationship, Nordhoff was not prepared to uphold Ganz's claim to be the spiritual father of the Volkswagen. He stated explicitly that for Volkswagen there was "no doubt at all" that the Volkswagen could be traced in its entirety "to the work of Professor Porsche and his staff."[54] Josef Ganz still greatly respected Ferdinand Porsche, but despite his ailing health he was determined to tell the world of his role in the creation of the Volkswagen. He probably did not have much time left. He had made only a slight recovery after his last heart attack. Financially too he was in such a bad way that he was dependent on gifts from his family in Switzerland. His seventy-one-year-old sister Margit, who was doing her best to support herself by teaching, did all she could to send Josef a hundred Swiss francs every month, a sum that was supplemented by his cousins, nephews, and nieces.[55] A woman friend in Melbourne brought him food from time to time.[56] He never did receive a pension from Volkswagen.

In telling his extraordinary life story to the media, Ganz was helped by a friend who passed on a collection of documents about him and his Volkswagen designs to the journalist Cyril Posthumus. A long article on the subject appeared in the British car magazine *Motor*. Although

Posthumus found it difficult to say exactly how much Ganz had contributed to the development of the Volkswagen, he concluded that he "ought certainly to be given full credit as a remarkable visionary," since it was he who "foresaw a design formula at least twelve years before it triggered off one of Germany's greatest technical-economical successes ever, and who tirelessly advocated its use until fortune turned against him."[57] In 1965, Ganz got in touch directly with the car magazine *Australian Motor Sports & Automobiles,* and the editors allocated a few pages to him in their June issue so he could tell his story. The item was announced on the cover, underlined and in big bright-yellow letters: "Volkswagen Shock! Hitler Stole My Design Says Melbourne Man!" Ganz praised Ferdinand Porsche, "whom I respect as a fine engineer," but claimed full credit for laying the basis for the Volkswagen project. "Apart from the position of the engine behind the axle," he concluded, "the Volkswagen of today principally is little different to the one I proposed in 1928. They haven't even changed the name!"[58]

Meanwhile, the old Maikäfer prototype of 1931, which marked the start of the development of the Volkswagen, had been languishing for years in the City Garage in Zürich, where Ganz had put it into storage during the war. His Tatra had been there too, and that had been scrapped. Now the Dutch owner of the garage was threatening to take possession of the May Bug, because of overdue rent on the space. Ganz had already received a payment demand from Switzerland, but in his extremely difficult situation in Australia he could only hope that someone would save his beloved Maikäfer from the wreckers.[59]

His long illness had prevented Ganz from supporting Volkswagen in its case against Tatra and from traveling to Germany as a witness, so the company negotiated a deal that was signed on August 12, 1965. To avoid further lengthy legal proceedings, Volkswagen paid Tatra one million Deutschmark for a license for the use of patent DRP 636633—one of the three patents for which Tatra had originally applied in December 1932 with the express purpose of thwarting Ganz's attempts to develop a Volkswagen.

Order of Merit

The articles published in Germany, Britain, and Australia about the work of Josef Ganz were not without result. In October 1965, the West German embassy in Australia contacted the Australian government to say that the Germans wanted to give Ganz the Bundesverdienstkreuz I. Klasse des Verdienstordes der Bundesrepublik Deutschland. To award this order of merit, the German embassy needed the official permission of the Australian government. It explained its reasons:

> It is in his capacity as editor of the magazine "*Motor-Kritik*" between 1928 and 1934 that Mr. Ganz took a keen interest in the development of a German "Volkswagen" and, together with other engineers like Professor Porsche, has greatly contributed to realizing this project. Furthermore, by promoting the idea of using rear engines, backbone platform chassis, and swing axle suspensions, he has greatly furthered the German automobile industry.[60]

Josef Ganz never received his order of merit. The Australian government permitted the awarding of foreign decorations only if the services they were intended to recognize had been rendered no more than five years before application was made. "Mr. Ganz would not appear to be entitled to receive either full or restricted permission to accept and wear this award," the Prime Minister's Department decided. "Mr. Ganz has not rendered any service to the German industry since 1934."[61] Ganz knew nothing of this German attempt to decorate him. He spent his days largely at home and was delighted by every letter he received from family, friends, and acquaintances in Europe. "My situation is not very cheerful," he wrote in February 1966 to Max Rauck of the Deutsches Museum in Munich. "This stupid heart...." He was disappointed to read in the media that the technical director of Tatra, Hans Ledwinka, was to be awarded the German order of merit, whereas he himself had not heard a word. Ganz wrote to Rauck expressing his regret that he could not continue their correspondence with carbon copies from his autobi-

PRIME MINISTER'S DEPARTMENT

TELEPHONE 71411
IN REPLY QUOTE 65/5431

25 NOV 1965

CANBERRA, A.C.T.

24 NOV 1965

The Secretary,
Department of External Affairs,
CANBERRA. A.C.T.

I refer to your memorandum 1535/25/17, concerning the proposed award of Officer's Cross of the Order of Merit of the Federal Republic of Germany to Mr. Josef Ganz of 12 Marine Place, St. Kilda, Victoria.

Under Regulations concerning the Acceptance and Wearing of Foreign Orders, Decorations and Medals by Persons not in the Service of the Crown, Mr. Ganz would not appear to be entitled to receive either full or restricted permission to accept and wear this award. Regulation 4 clearly provides that permission will not be given where the services to be recognised have been wholly rendered more than five years before the question of eligibility for permission is raised. Mr. Ganz has not rendered any services to the German motor industry since 1934.

(E. J. Bunting)
Secretary

Mr Prettman

Decision letter from the Australian Prime Minister's Department, 1965 ography; he was too weak. Nevertheless, he still followed the motoring news and was extremely impressed by the well-designed and well-built cars the Japanese were putting onto the market. He called the new compact Honda 600 "a work of art."[62] Although new models from Japanese, European, and American manufacturers represented serious competition for the VW Beetle, Volkswagen's director Heinrich Nordhoff stuck firmly to a sales strategy based on the company's one strong model.

Ganz rarely left his apartment. He spent most of his time lying on the bed or the sofa, watching television. The misery of the world that passed by on his black-and-white screen made him quite somber, and sometimes he lay awake all night. After everything he had experienced, it saddened him that people were unable to get along. In Germany, the West and the Russians, both armed to the teeth, were still engaged in their Cold War.[63] When he felt strong enough to write or type, he tried to correspond a little with those close to him, but they were all thousands of miles away, including his old love and life's companion Madeleine Paqué, his sister Margit, and his former colleagues and good friends Georg Ising and Frank Martin. They continued to send him clippings from German car magazines and newspapers whenever they spotted anything about the history of the Volkswagen. One article declared that technical director Hans Ledwinka of Tatra was the spiritual father of the Volkswagen, another that Adolf Hitler had thought up the nickname Maikäfer. "And so they go on eagerly lying," Ganz wrote in November 1966 in an emotional letter to his sister Margit in Switzerland. He would have liked nothing better than to visit her and the rest of his close relatives. He longed to see his nephew Hugo, who had become an engineer just like him, and to watch his great-niece Maja grow up. "My dearest, how good it would be if we could chat together," Josef Ganz wrote to his sister, begging her to write back quickly. A modern jet, the Boeing 707, could make the journey from Australia to Switzerland in less than twenty-four hours, but Ganz would not have been up to it even if he had been able to find the money for a ticket.[64] It remained no more than a dream; he would never see his family or old friends again. Georg Ising, who had meant so much to him for so many years, died a short while later, in January 1967 at the age of eighty-six.

Ganz felt his final days ticking away as he lay in his lonely sickbed. Hardly more than six months after Georg Ising, on July 26, 1967, Josef Ganz died, a few weeks after his sixty-ninth birthday, in St. Kilda, from the effects of another heart attack. A few days later, on July 31, he was cremated in a private ceremony at Springvale Crematorium, a beautiful, quiet cemetery with its own crematorium at the edge of the little coastal suburb of Melbourne. No trace of him was left, no urn and no commemorative plaque.

Sales Success

The extensive archive with countless pieces of evidence that Josef Ganz left when he died was largely lost. His apartment was cleared by his good friend and lawyer John Kinnear, who gathered up and kept only the photographs and the microfilm archives. Director Heinrich Nord hoff of Volkswagen either had no interest in taking care of the files or was not in a position to do so. Since the summer of 1967, he had been struggling with heart problems of his own, and he died less than a year after Ganz, on April 12, 1968, also aged sixty-nine. After Nordhoff's death, Josef Ganz was almost entirely forgotten. In 1972, when journalist John Winding-Sorensen of the British magazine *Car* contacted both GM Holden in Australia and Volkswagen in Germany in the hope of finding out more about Josef Ganz, GM Holden merely sent him a standard letter with a few car brochures, and Volkswagen failed to respond at all. The only historical source he was able to find was the article Ganz had written for *Australian Motor Sports & Automobiles* seven years earlier. In his own article, Winding-Sorensen observed that it was "a bit irritating" that the Germans stuck so persistently to the claim that the Volkswagen was a unique design "executed by this one genius, Ferdinand Porsche." The journalist believed it was "a true German myth which is kept alive at all costs." The documents relating to Josef Ganz and Porsche's first prototypes for Zündapp were "more heavily protected than recent intelligence reports." Winding-Sorensen concluded that no matter what,

Ganz deserved "some kind of recognition," given that he was one of the very best auto journalists and that "as an automobile engineer he was in the same class as Porsche, Rumpler and Nibel."[65]

In the twenty years that Heinrich Nordhoff was in charge, Volkswagen had relied primarily on a single model, the Beetle. Its one-model policy was relinquished only after Nordhoff died, and despite the introduction in 1974 of its successor the Volkswagen Golf—a contrasting concept with a front-mounted, water-cooled engine and front-wheel drive—the popular VW Beetle would remain in production in Germany until 1978. Production continued for many years after that in Brazil and Mexico. Because of the still-unmatched popularity of the Beetle, Volkswagen produced a New Beetle in 1998, a car based on the Golf 4 but with bodywork inspired by its original Bug. The very last authentic Volkswagen Beetle rolled off the production line in Mexico in 2003. In total, more than twenty-one million had been built—an unprecedented success story that began more than seventy years earlier with a little Maikäfer from Frankfurt.

ACKNOWLEDGMENTS

During my five years of research on the subject of Josef Ganz, I have come into contact with countless private individuals as well as the staffs of archives and museums in Europe, Australia, and the United States. My heartfelt thanks are due to all of them for their help in tracking down archival material and their support in piecing together the story as thoroughly and comprehensively as possible. I would like to thank the following by name:

—David Allonby of Reprodyne, for reconstructing design drawings by Josef Ganz in digital form.
—L. Scott Bailey, founder of *Automobile Quarterly*, for making unique illustrations available.
—Dieter Dressel, for access to his private archive.
—Gaye Eksen, for help with research and support during the writing of this book.
—René van Praag and Astrid Bosch of RVP Publishers, for their enthusiasm and for having the confidence to publish the book in English.
—Manfred Grieger of the Corporate History Department, Volkswagen AG, for access to the company archives.

—Gerry Harant, for assistance in research and for sharing his memories of Josef Ganz.
—Otfried Jaus, for access to his private archive.
—Dieter Klüpfel, for a wealth of photographic material and for sharing his many memories of Josef Ganz.
—John Lavery, for access to his private archive.
—Michael Lenarz of the Jewish Museum, City of Frankfurt am Main, for assistance and support during my research.
—Richard Neukom, for access to his private archive.
—Wolfgang Rabus of Mercedes-Benz Archives & Collections, Daimler AG, for access to the company archives.
—Ton and Rini Schilperoord, for assistance and support during my research.
—Maja von Tolnai, for sharing her family memories and providing photographic material.
—Andreas Schmid and Lorenz Schmid, for assistance and support during my research.
—Meira Shacham, for access to her private archive.
—Patricia A. Billingsley, for access to her private archive.
—Aimée Warmerdam, for substantive advice and for the editing of several early chapters.
—Liz Waters, for her excellent translation work from the original Dutch text.

CREDITS

Most of the illustrations in this book are from Josef Ganz's personal archive. The photographs were in most cases taken by Josef Ganz himself. Other sources that provided illustrations are listed below.

David Allonby from Reprodyne: fourth photo insert, p. 16 computer renderings, center right and both at bottom; fifth photo insert, p. 5, both computer renderings at bottom.

Patricia A. Billingsley, private collection: p. 204.

b p k Photo Agency, Berlin: fifth photo insert, p. 11 and p. 33 top.

Estate of Philip Cummings: fifth photo insert, p. 19 bottom.

Hessisches Hauptstaatsarchiv, Wiesbaden: p. 10, p. 128, p. 142, pp. 202-203.

Mercedes-Benz Archives & Collections, Daimler AG, Stuttgart: p. 73, first photo insert, p. 7, both pictures at bottom, page 8 top right, all photographs on p. 9, p. 12 top, p. 31 top; second photo insert, p. 10 center right and bottom right, p. 11 top and bottom; fourth photo insert, p. 3 bottom, p. 16 top and center left; fifth photo insert, pp. 8-9.

Museum Industriekultur, Nürnberg: second photo insert, p. 12 top.

National Aeronautics and Space Administration (NASA): first photo insert, p. 5 top right.

National Archives of Australia: p. 254.

Dr. Ing. h.c. F. Porsche Aktiengesellschaft: second photo insert, p. 13, all images except top; fourth photo insert, p. 5 top; fifth photo insert, p. 35 top.

Corporate History Department, Volkswagen AG, Wolfsburg: fifth photo insert, pp. 32-35, all photos except p. 33 top, p. 34, and p. 35 top.

For the illustrations used in the book every effort has been made to trace the holders of copyright and to acknowledge the permission of authors and publishers where necessary. If we have inadvertently failed in this aim, we will be pleased to correct any omissions in future editions of this book.

NOTES

Prologue

1 W. H. Millgate, "German 'Baby' Car Has Engine Mounted in Rear," in *The Detroit News*, April 23, 1933.

Chapter One: From Tropfenwagen to Volkswagen

1 Josef Ganz, "Ein Denkmal für den Ersterfinder des Benzinautomobils," in *Motor-Kritik 12*, no. 22, mid-November 1932, pp. 21-22.

2 Letter from Ferdinand Fuchs to the district court in Zürich, June 9, 1944. [author's archive]

3 Ibid.

4 "Porträts zur Tagesgeschichte," *Stadt Gottes, Illustrierte Zeitschrift für das katholische Volk*, 1910, p. 377.

5 Jakob Feldhammer's biography (work in progress), based on his memoirs (written in 1941-1944). [archive of Meira Shacham, meira.shacham@gmail.com]

6 Ibid.

7 Letter from Ferdinand Fuchs to the district court in Zürich, June 9, 1944. [author's archive]

8 Jakob Feldhammer's Biography (work in progress) based on his memoirs (written in 1941-1944). [archive of Meira Shacham, meira.shacham@gmail.com]

9 Letter from Ferdinand Fuchs to the district court in Zürich, June 9, 1944. [author's archive]

10 Testimonial on Josef Ganz by the directors of Chemische Fabriken Worms AG, December 1, 1920. [author's archive]

11 Josef Ganz, "Als Pfadfinder auf dem Wege vom Benz-Komfortable zum MB 130 Heckwagen. Ein stuck technischer Geschichte, an dem ich teilhatte," in *Motor-Kritik 14*, no. 1, early January 1934, pp. 3-5.

12 Testimonial on Josef Ganz by the directors of Chemische Fabriken Worms AG, December 1, 1920. [author's archive]

13 Patent DRP [DE] 383063 ("Wendeschraube"), property of Josef Ganz, application dated September 13, 1921, granted on October 9, 1923.

14 Josef Ganz, "Der bei uns vergessene Motorradwagen," in *Klein-Motor-Sport 11*, no. 8, late April 1922, pp. 119-120.

15 Josef Ganz, "Als Pfadfinder auf dem Wege vom Benz-Komfortable zum MB 130 Heckwagen. Ein stuck technischer Geschichte, an dem ich teilhatte," in *Motor-Kritik 14*, no. 1, early January 1934, pp. 3-5.

16 Ibid.

Chapter Two: A New Calling

1 Patent DRP [DE] 130895 ("Untergestell für Motorfahrzeuge"), property of Dr. Georg Klingenberg, application dated July 16, 1901, granted on May 16, 1902.

2 Letter to Josef Ganz dated February 15, 1923, from the Berlin Patent Office with patent application ("Druckausgleicher für Ansaugleitungen von Verbrennungsmotoren"). [author's archive]

3 Letter from Ferdinand Fuchs to the district court in Zürich, June 9, 1944. [author's archive]

4 Josef Ganz, "Kleinstwagen," in *Deutsche Motorsport Zeitung*, June 8, 1924.

5 Patent CH 105267 ("Spielzeug"), property of Josef Ganz, application dated May 17, 1923, granted on June 16, 1924.

6 Patent DRP [DE] 417427 ("Magnetelektrische Getriebe"), property of Josef Ganz, application dated January 8, 1924, granted on August 12, 1925.
7 Josef Ganz, "Ist im Kraftfahrzeugbau der Gipfelpunkt erreicht?," in *Motor und Sport*, nos. 12, 13, & 26, 1926.
8 S. Lubinski, "Die heroischen 'Narren': Pioniere der Kraftfahrt," in *Das Auto Motor und Sport*, no. 26, December 22, 1956, pp. 58-61, 63.
9 Josef Ganz, "Schaden an Körper und Gesundheit," September 19, 1966. [Hessisches Hauptstaatsarchiv, Wiesbaden]
10 Josef Ganz, "Kritische Betrachtungen über Konstruktions-Details am Hanomag," in *Klein-Motor-Sport 8*, no. 5, mid-March 1928, pp. 97-102.
11 Josef Ganz, "Wie wird der zukünftige Kleinwagen aussehen?," in *Klein-Motor-Sport 8*, no. 24, late December 1928, pp. 528-536.
12 Josef Ganz, "Ist im Kraftfahrzeugbau der Gipfelpunkt erreicht?," in *Motor und Sport*, nos. 12, 13, & 26, 1926.
13 Josef Ganz, "Für den deutschen Volkswagen," in *Klein-Motor-Sport 8*, no. 1, mid-January 1928, pp. 1-2.
14 Josef Ganz, "Nicht einmal den kleinen Finger gibt man uns ganz!," in *Klein-Motor-Sport 8*, no. 1, mid-January 1928, p. 12.
15 Front page of *Klein-Motor-Sport 8*, no. 10, late May 1928.
16 Letter from G. Schlesinger of the University of Brussels to Ferdinand Fuchs, April 15, 1936. [author's archive]
17 Josef Ganz, "Der Röhrwagen, ein Zeichen des Wiedererwachens der deutschen Automobilbaukunst," in *Klein-Motor-Sport 8*, no. 12, late June 1928, pp. 239-251.
18 Josef Ganz, "Mit dem kleinen Dixi kreuz und quer über Schweizer Alpenpässe," in *Klein-Motor-Sport 8*, no. 16, late August 1928, pp. 329-346.
19 Letter from Josef Ganz to Max J. B. Rauck, custodian of the Deutsches Museum in Munich, March 2, 1962. [author's archive]
20 Josef Ganz, "Neue 'Tatra'-Modelle," in *Klein-Motor-Sport 8*, no. 8, late April 1928, pp. 160-161.
21 Josef Ganz, "Bilanz der 'modernen' Automobilkonstruktion. Vor allem: Bautet Schwingachsen ein!," in *Klein-Motor-Sport 8*, no. 19, mid-October 1928, pp. 395-399.
22 Josef Ganz, "Keine Wagen mehr ohne Schwingachsen!," in *Klein-Motor-Sport 8*, no. 19, mid-October 1928, pp. 399-409.
23 OW [Oskar Weller], "Kleinwagen—nicht Miniatur-Auto!," in *Motor-Kritik 9*, no. 1, mid-January 1929, pp. 12-13.
24 Josef Ganz, "Ein Anonymus!," in *Klein-Motor-Sport* 8, no. 21, mid-November 1928, pp. 458-459.

Chapter Three: *Motor-Kritik*

1 *Motor-Kritik 9*, no. 6, late March 1929, pp. 138-139.
2 Contract between H. Bechhold Verlagsbuchhandlung and Josef Ganz, February 27, 1929. [Hessisches Hauptstaatsarchiv, Wiesbaden]
3 Announcement in *Motor-Kritik 9*, no. 1, mid-January 1929, p. 11.
4 Josef Ganz, "Das Modell zum Europa-Wagen ist da!," in *Motor-Kritik 9*, no. 1, mid-January 1929, pp. 14-16.
5 "Klein-Motor-Sport—Motor-Kritik," in *Motor-Kritik 9*, no. 1, mid-January 1929, p. 11.
6 "Rationalisierung der Fachpresse," in *Motor-Kritik 9*, no. 21, mid-November 1929, p. 453.
7 Josef Ganz, "Die RDA-Blockade gegen die Fachpresse," in *Motor-Kritik 11*, no. 2, mid-January 1931.
8 Armin Drechsel, "Etwas über das Kreiselautomobil," in *Motor-Kritik 10*, no. 8, mid-April 1930, pp. 169-174.
9 Béla Barényi, "Konstruktion eines Einzelträger-Chassis," in *Motor-Kritik 9*, no. 11, mid-June 1929, pp. 240-241.
10 "Warum wir keine Schwingachsen bauen," in *Stoewer Magazin*, no. 3, March 1929.
11 "Kaleidoskop" column, in *Motor-Kritik 9*, no. 5, mid-March 1929, p. 116.
12 Oskar Weller, "Was uns nicht gefällt," in *Motor-Kritik 9*, no. 17, mid-September 1929, pp. 369-370.
13 "Brennabor-Juwel und die 'Motor-Kritik,'" in *Brennabor-Roland*, no. 5, October 1929.
14 "Motor-Kritik-Wettbewerb," in *Motor-Kritik 9*, no. 18, late September 1929, p. 385.
15 Cover of *Motor-Kritik 9*, no. 20, late October 1929.
16 Cover of *Motor-Kritik 9*, no. 13, mid-July 1929.
17 Oskar Weller, "Hallo! Wir fahren nach Paris," in *Motor-Kritik 9*, no. 19, mid-October 1929, pp. 407-409.
18 Oskar Weller, "Man sieht... Man hört... (Impressionen vom Pariser Salon de l'Automobile)," in *Motor-Kritik 9*, no. 19, mid-October 1929, pp. 414-418.
19 Patent DRP [DE] 528643 ("Fahrgestell für Kraftfahrzeuge"), property of Josef Ganz, application dated June 18, 1929, granted on June 18, 1931.
20 Letter from Josef Ganz to Max J. B. Rauck, custodian of the Deutsches Museum in Munich, March 2, 1962. [author's archive]
21 Josef Ganz, "Röhr-Test," in *Motor-Kritik 10*, no. 8, mid-April 1930, pp. 155-159.
22 Cover of *Motor-Kritik 10*, no. 6, early March 1930.
23 Advertisement by Daimler-Benz in *ADAC-Motorwelt*, no. 10, March 7, 1930.
24 Josef Ganz, "Mercedes-Ethik," in *Motor-Kritik 10*, no. 6, mid-March 1930, pp. 115-120.
25 Letter from Josef Ganz to Max J. B. Rauck, custodian of the Deutsches Museum in Munich, February 5, 1962. [Corporate History Department, Volkswagen AG, Wolfsburg]
26 Agreement between H. Bechhold Verlagsbuchhandlung and Daimler-Benz, April 2, 1930. [author's archive]

Chapter Four: The Bug Is Born

1 Letter from Carl Hahn of DKW to Josef Ganz, June 26, 1930. [Staatsarchiv des Kantons Zürich]
2 Josef Ganz, "Welchen Wagen soll ich mir anschaffen?," in *Motor-Kritik 10*, no. 12, mid-June 1930, pp. 256-258.
3 "Kaleidoskop" column, in *Motor-Kritik 10*, no. 14, mid-July 1930, pp. 289-293.
4 Josef Ganz, "Das Ende der Tropfenwagenidee—oder ihre Auferstehung?," in *Motor-Kritik 10*, no. 13, early June 1930, pp. 263-269.
5 Letter from Willy Bendit of Ardie to Josef Ganz, March 18, 1932. [author's archive]
6 Josef Ganz, "Citroën-Test," in *Motor-Kritik 10*, no. 16, mid-August 1930, pp. 334-338.
7 Josef Ganz, "How I Invented the Volkswagen," in *Australian Motor Sports & Automobiles*, June 1965, pp. 6-10.
8 Walter Ostwald, "Fahrmaschinenversuche. Drei oder vier Räder?," in *Allgemeine Automobil Zeitung* (*AAZ*), fall issue 1930, p. 11.
9 Letter from Josef Ganz to Max J. B. Rauck, curator of the Deutsches Museum in Munich, March 2, 1962. [author's archive]
10 Josef Ganz, "Der Ardie-Versuchs-Kleinstwagen 1930," in *Motor-Kritik 10*, no. 18, mid-September 1930, pp. 390-391.
11 Letter from Ferdinand Fuchs to the district court in Zürich, June 9, 1944. [author's archive]
12 Letter from Willy Bendit of Ardie to Josef Ganz, March 18, 1932. [author's archive]
13 Josef Ganz, "Die Führerscheinfreiheit für Kraftvierräder hintertrieben," in *Motor-Kritik 11*, no. 3, early February 1931, p. 45.
14 Josef Ganz, "Eine deutsche Karosserie der Sachlichkeit," in *Motor-Kritik 10*, no. 20, mid-October 1930, pp. 450-451.
15 Letter from Josef Ganz to Hans Nibel of Daimler-Benz, December 2, 1930. [author's archive]
16 Patent DRP [DE] 585116 ("Lenkvorrichtung für einzeln lenkbare Lenkräder von Fahrzeugen, insbesondere Kraftfahrzeugen"), property of Josef Ganz, application dated February 17, 1931, granted on September 14, 1933.
17 Patent DRP [DE] 728723 ("Anordnung für Fußhebel, insbesondere für Kraftfahrzeuge"), property of Josef Ganz, application dated October 22, 1931, granted on October 29, 1942.
18 Letter from Josef Ganz to Hans Nibel of Daimler-Benz, December 2, 1930. [author's archive]
19 Letter from Hans Nibel of Daimler-Benz to Josef Ganz, December 16, 1930. [author's archive]
20 Letter from Emil Georg von Stauss of Deutsche Bank to Ferdinand Porsche, July 4, 1930. [Bundesarchiv, Berlin]
21 Letter from Ferdinand Porsche to Emil Georg von Stauss of Deutsche Bank, June 25, 1930. [Bundesarchiv, Berlin]
22 Letter from Von Lewinski of Deutsche Bank to Emil Georg von Stauss of Deutsche Bank, January 9, 1931. [Bundesarchiv, Berlin]
23 Letter from Josef Ganz to Hans Nibel of Daimler-Benz, December 17, 1930. [author's archive]

Chapter Five: Sabotaged Bug

1 Gedenkschrift betr. Kleinstwagen by Josef Ganz, July 25, 1931. [author's archive]
2 Oskar Weller, "Ausstellungs-Nachlese. Die Mark hat hundert Pfennige...," in *Motor-Kritik 11*, no. 6, mid-March 1931, pp. 119-123.
3 Josef Ganz, "Hurra, der erste Großserien-Kleinstwagen ist da!," in *Motor-Kritik 11*, no. 3, early February, 1931, pp. 42-43.
4 Josef Ganz, "Eine Seite 'Maikäfer,'" in *Motor-Kritik 11*, no. 21, early November 1931, pp. 472-473.
5 Janus (Paul G. Ehrhardt), "Luftwiderstand und Auto. Komödie der Irrungen mit ernsthaftem happy end," in *Motor-Kritik 11*, no. 10, mid-May 1931, pp. 211-212.
6 Draft contract between Zündapp and "B" (Fidelis Böhler), date unknown (probably 1931). [Deutsches Technikmuseum, Berlin]
7 Josef Ganz, "Ein Zündapp-Hanomag?," in *Motor-Kritik 11*, no. 10, mid-May 1931, pp. 218-219.
8 "Kaleidoskop" column, in *Motor-Kritik 11*, no. 19, early October 1931, pp. 426-427.
9 Janus (Paul G. Ehrhardt), "Quod erat demonstrandum," in *Motor-Kritik 11*, no. 14, mid-July 1931, pp. 299-301.
10 From a statement to the court by Georg Ising, February 1934. [author's archive]
11 Letter from Paul G. Ehrhardt to Josef Ganz, November 16, 1931. [author's archive]
12 Josef Ganz, "Die Schwingachse des Kleinen," in *Motor-Kritik 11*, no. 3, early February 1931, pp. 41-42.
13 Letter from the directors of BMW to Josef Ganz, July 20, 1931. [Hessisches Hauptstaatsarchiv, Wiesbaden]

Chapter Six: Intrigues in Paris

1 *Automobiltechnisches Handbuch—Handbuch der Automobiltechnischen Gesellschaft E.V.*, M. Krayn Technischer Verlag, Berlin, 1931.
2 Letter from Alex Taub, Membership Committee Chairman of the Society of Automotive Engineers (SAE) to Josef Ganz, June 22, 1931. [author's archive]
3 Letter from Josef Ganz to Max J. B. Rauck, custodian of the Deutsches Museum in Munich, March 2, 1962. [author's archive]
4 Josef Ganz, "Probefahrt-Bericht über den Mercedes, Typ 170," in *Motor-Kritik 11*, no. 19, early October 1931, pp. 416-419.
5 Interview with Gerry Harant, an ex-colleague and friend of Josef Ganz in Australia, July 2008. [author's archive]

6 Kraemer, *Neue Pariser Zeitung*, October 12, 1931.
7 From a statement to the court by Georg Ising, February 1934. [author's archive]
8 Letter from director Ernst Decker of Neuen Röhr-Werke to Josef Ganz, August 24, 1931. [author's archive]
9 Letter from Hermann Klee, member of the board at Wanderer-Werke AG to Josef Ganz, January 28, 1932. [author's archive]
10 Paul Friedmann "Apologie wegen Paris," with a reaction by Frank Arnau, in *Motor-Kritik 11*, no. 21, early November 1931, pp. 478-484.
11 Letter from Knappe to Josef Ganz, November 17, 1931. [author's archive]
12 Josef Ganz, "Mercedes Typ 170," *Motor-Kritik 11*, no. 19, early October 1931.
13 Letter from Wilhelm Kissel of Daimler-Benz to Josef Ganz, October 2, 1931. [author's archive]
14 Letter from Ernst Neumann-Neander to Josef Ganz, October 1, 1931. [author's archive]
15 Letter from Willy Bendit of Ardie to Josef Ganz, October 6, 1931. [author's archive]
16 Frank Arnau, "Mercedes-Benz 'Typ 170'—der neue Wagen der Untertürkheimer Werke," in *Allgemeine Automobilzeitung*, no. 40, October 3, 1931.
17 Letter from Paul G. Ehrhardt to Josef Ganz, November 16, 1931. [author's archive]
18 Josef Ganz, "Mercedes Typ 170," in *Motor-Kritik 11*, no. 19, early October 1931.
19 Paul Jaray, in *Motor-Kritik 11*, no. 19, early October 1931.
20 Letter from Josef Ganz to technical director Hans Nibel of Daimler-Benz, November 19, 1931. [author's archive]
21 Patent DRP [DE] 642192 ("Kraftfahrzeug mit vier unabhängig abgefederten Rädern"), property of Daimler-Benz, application dated April 22, 1932, granted on February 11, 1937.
22 Letter from the directors of Daimler-Benz to Josef Ganz, November 19, 1931. [author's archive]
23 Letter from technical director Hans Nibel of Daimler-Benz to Josef Ganz, November 28, 1931. [author's archive]
24 Contract between Zündapp and Dr.-Ing. h.c. Ferdinand Porsche GmbH, September 28, 1931. [Deutsches Technikmuseum, Berlin]
25 Letter from Ferdinand Porsche to director Fritz Neumeyer of Zündapp, September 19, 1931. [Deutsches Technikmuseum, Berlin]
26 Griffith Borgeson, "In the Name of the People—the True Origins of the VW Beetle," in *Automobile Quarterly*, no. 4, 4th quarter, 1980, pp. 340-361.
27 Contract between Zündapp and Dr.-Ing. h.c. Ferdinand Porsche GmbH, September 28, 1931. [Deutsches Technikmuseum, Berlin]
28 Cover of *Motor-Kritik* 11, no. 21, early November, 1931.
29 Georg Ising, "Deutsche Rennwagen 1934," *Motor-Kritik 14*, no. 3, early February 1934, pp. 60-61.
30 Letter from Armin Drechsel to Ferdinand Fuchs, April 20, 1944. [author's archive]
31 Minutes of a meeting at Zündapp, November 6, 1931. [Deutsches Technikmuseum, Berlin]

Chapter Seven: Nazi Smear Campaign

1 "Kaleidoskop" column, in *Motor-Kritik 11*, no. 22, mid-November 1931, p. 508.
2 "Kaleidoskop" column, in *Motor-Kritik 12*, no. 4, mid-February 1932, p. 93.
3 Josef Ganz, "Eine Seite 'Maikäfer,'" in *Motor-Kritik 11*, no. 21, early November 1931, pp. 472-473.
4 Anon., "'Motor-Kritik,'" in *Die Nationale Front—Kampfblatt für Deutsche Politik und Deutsche Wirtschaft*, no. 12, November 1-15, 1931, p. 10.
5 "Kaleidoskop" column, in *Motor-Kritik 11*, no. 22, mid-November 1931, p. 508.
6 Letter from Josef Ganz to the directors of Daimler-Benz, November 8, 1931. [author's archive]
7 Letter from the directors of Daimler-Benz to Josef Ganz, November 10, 1931. [author's archive]
8 Josef Ganz, "Wettervorhersage im Automobilbau," in *Motor-Kritik 11*, no. 10, mid-May 1931, pp. 207-210.
9 Wolfgang van Lengerke, "Das Auto von Morgen?," in *Motor und Sport 8*, no. 46, November 15, 1931, pp. 9-10.
10 Josef Ganz, "Eine deutsche Karosserie der Sachlichkeit," in *Motor-Kritik 10*, no. 20, mid-October 1930, pp. 450-451.
11 Wolfgang von Lengerke, "Das Auto von Morgen?," in *Motor und Sport 8*, no. 49, December 6, 1931, pp. 13-14.
12 Anon., "Konferenz! Bitte nicht stören!—Traum einer Neujahrsnacht," in *Motor und Sport 8*, no. 52, Christmas 1931, pp. 9-13.
13 "Kaleidoskop" column, in *Motor-Kritik 12*, no. 1, early January 1932, p. 22.
14 Anon., "'Motor-Kritik!,'" in *Die Nationale Front—Kampfblatt für Deutsche Politik und Deutsche Wirtschaft*, no. 14, Christmas 1931, p. 12.
15 Appeal in *Motor-Kritik 12*, no. 1, early January 1932, p. 7.
16 Josef Ganz, "Zwölfter Jahrgang 'M-K,'" in *Motor-Kritik 12*, no. 1, early January 1932.
17 Letter from Hermann Klee, member of the board at Wanderer-Werke AG, to Josef Ganz, January 28, 1932. [author's archive]
18 Letter from Walter Schneider to Josef Ganz, February 1, 1932. [author's archive]
19 Letter from W. Stoekicht to Josef Ganz, February 6, 1932. [author's archive]
20 Letter from Josef Ganz to Hans Nibel of Daimler-Benz, March 14, 1932. [author's archive]
21 Letter from Hans Nibel of Daimler-Benz to Josef Ganz, March 17, 1932. [author's archive]
22 Letter from Willy Bendit of Ardie to Josef Ganz, March 18, 1932. [author's archive]
23 Letter from K. D. von Oertzen, member of the board at Wanderer-Werke, to Josef Ganz, March 18, 1932. [author's archive]

24 "Kaleidoskop" column, in *Motor-Kritik 12*, no. 8, mid-April 1932, p. 173.
25 Letter from Ferdinand Fuchs to the district court in Zürich, June 9, 1944. [author's archive]
26 Statement by Hans Köth and Robert Allmers, March 26, 1932. [author's archive]
27 "Kaleidoskop" column, in *Motor-Kritik 12*, no. 10, mid-May 1932, p. 240.

Chapter Eight: The May Bug Becomes the Superior

1 Josef Ganz, "Der neue BMW," in *Motor-Kritik 12*, no. 7, early April 1932.
2 Béla Barényi, "'Frontdrive' 'Standard-' 'Streamline' oder Heckmotor Rohrchassis Stromlinie—Experímentum crucís," in *Motor-Kritik 12*, no. 15, early August 1932, pp. 348-353.
3 Letter from Hans Nibel of Daimler-Benz to Josef Ganz, June 17, 1932. [author's archive]
4 Patent DRP [DE] 587409 ("Antriebsblock für Kraftfahrzeuge"), property of Josef Ganz, application dated May 8, 1932, granted on October 19, 1933.
5 Patent DRP [DE] 83749 ("Achsanordnung, insbesondere für Kraftfahrzeuge"), property of Josef Ganz, application dated May 8, 1932, granted on August 24, 1933.
6 Patent DRP [DE] 576741 ("Einrichtung zur Belüftung des Motorraumes bei Fahrzeugen mit Heckmotor"), property of Josef Ganz, application dated May 14, 1932, granted on May 4, 1933.
7 Gebrauchsmuster DRGM 1254347 ("Schaltvorrichtung"), property of Josef Ganz.
8 Gebrauchsmuster DRGM 1253035 ("Anordnung d.el. Kabel an Kraftfahrzeuge"), property of Josef Ganz.
9 Josef Ganz, "1200 ccm Tatra-Vierzylinder-Kurztest," in *Motor-Kritik 12*, no. 13, early July 1932, pp. 305-308.
10 Statement by Wilhelm Gutbrod, June 17, 1941. [Hessisches Hauptstaatsarchiv, Wiesbaden]
11 Josef Ganz, "'Ein Laune des Schicksals: Der erste 'Maikäfer', ein 'Standard'-Wagen," in *Motor-Kritik 13*, no. 4, mid-February 1933, pp. 102-103.
12 From a statement to the court by Josef Ganz, February 3, 1937. [author's archive]
13 Patent DRP [DE] 569670 ("Anordnung des als Sternmotor ausgebildeten Antriebsmotors an Kraftfahrzeugen, insbesondere an solchen mit schwingbarer Halbachsen und einem mittleren Tragkörper als Wagenrahmen"), property of Dr.-Ing. h.c. F. Porsche GmbH, application dated November 25, 1931, granted on January 19, 1933.
14 "Kaleidoskop" column, in *Motor-Kritik 12*, no. 18, mid-September 1932, p. 427.
15 Report in *Motor-Kritik 13*, no. 7, early April 1933, p. 191.
16 "Kaleidoskop" column, in *Motor-Kritik 12*, no. 16, mid-August 1932, pp. 379-380.
17 "Ketzereien gegen den Heckmotor! Eine Konstruktion wird preisgegeben!," in *Motor-Kritik 12*, no. 17, early September 1932, pp. 385-390.
18 Ibid.
19 Béla Barényi, "'Frontdrive' 'Standard-' 'Streamline' oder Heckmotor Rohrchassis Stromlinie—Experímentum crucís," in *Motor-Kritik* no. 15, early August 1932, pp. 348-353.
20 Patent DRP [DE] 423116 ("Kraftwagen mit aus einem Hohlträger bestehendem Fahrzeuggestell"), property of Arnold Seidel, application dated December 15, 1923, granted on December 19, 1925.
21 Patent DRP [DE] 448406 ("Fahrgestell für Vierradkraftwagen"), property of Paul G. Ehrhardt, application dated December 6, 1925, granted on August 12, 1927.
22 Josef Ganz, "Stehen dem Bau von Kraftwagen mit Mittelträgerrahmen Schutzrechte entgegen?," in *Motor-Kritik 12*, no. 19, early October 1932, pp. 436-443.
23 Patent DRP [DE] 559488 ("Wagenkasten für Kraftfahrzeuge mit in sich steifem Boden"), property of BMW, application dated July 24, 1931, granted on September 8, 1932. Patent DRP [DE] 559689 ("Fahrgestellrahmen für Kraftfahrzeuge"), property of BMW, application dated July 24, 1931, granted on September 8, 1932.
24 "BMW läßt das Gericht sprechen," in *Motor-Kritik 14*, no. 1, early January 1934, p. 18.
25 "Kaleidoskop" column, in *Motor-Kritik 11*, no. 19, early October 1931, p. 427.
26 Verdict of the district court in Frankfurt am Main in the case of BMW vs. Paul G. Ehrhardt, November 30, 1933. [author's archive]
27 Letter from director L. Dürr of Luftschiffbau Zeppelin to Paul Jaray, March 22, 1924. [author's archive]
28 Patent application T 39 910 II/360-37 ("Achsanordnung, insbesondere für Kraftfahrzeuge") by Tatra, dated December 1932, not granted.
29 Patent application T 39 261 II/630-38 ("Anordnung der schwingenden Halbachsen an einem rohrförmigen Tragkörper, insbesondere für Kraftfahrzeuge") by Tatra, dated December 1932, not granted.
30 Patent DRP [DE] 636633 ("Lagerung eines Antriebsblockes an Kraftfahrzeugen, insbesondere mit einem z. B. rohrfoermigen Mittelrahmen"), property of Tatra, application dated December 18, 1932, granted on January 7, 1937.
31 Letter from Josef Ganz to director Heinrich Nordhoff of Volkswagen, October 27, 1961. [Corporate History Department, Volkswagen AG, Wolfsburg]
32 Letter from Josef Ganz to Max J. B. Rauck, custodian of the Deutsches Museum in Munich, March 2, 1962. [author's archive]
33 Anon., "Patente-Pioniere-Piraten," in *Motor und Sport 9*, no. 51, December 1932, pp. 8-9.
34 "Kaleidoskop" column, in *Motor-Kritik* no. 1, early January 1933.
35 Contract between Josef Ganz and Daimler-Benz, January 20, 1933. [author's archive]

36 Letter from Hans Nibel of Daimler-Benz to Josef Ganz, February 3, 1933. [author's archive]

Chapter Nine: Presentation for Hitler

1 Josef Ganz, "Der Berliner Ausstellung entgegen," in *Motor-Kritik 13*, no. 4, mid-February 1933, pp. 79-86.
2 Walter Ostwald, "Heckwagen mit Sternmotor. Ein neukonstruktion von Dr. Porsche. Vortest," in *Motor-Kritik 13*, no. 4, mid-February 1933, pp. 104, 106, 108, & 110.
3 Josef Ganz, "How I Invented the Volkswagen," in *Australian Motor Sports & Automobiles*, June 1965, pp. 6-10.
4 Josef Ganz, "Die Ausstellung der Ausstellungen: Berlin 1933," in *Motor-Kritik 13*, no. 5, early March 1933, pp. 117-137.
5 Ibid.
6 Josef Ganz, "How I Invented the Volkswagen," in *Australian Motor Sports & Automobiles*, June 1965, pp. 6-10.
7 Georg Ising, "Der verhinderte Volkswagen," in *Motor Rundschau 1*, no. 5, March 1, 1947, pp. 66-67.
8 Josef Ganz, "Ein Laune des Schicksals: Der erste 'Maikäfer,' ein 'Standard'-Wagen," in *Motor-Kritik 13*, no. 4, mid-February 1933, pp. 102-103.
9 Wilhelm Gutbrod, "Die Maikäfer-Komödie," in *Motor-Kritik 13*, no. 6, March 1933, pp. 166-167.
10 Witness statement by General Franz Halder during the case against Hermann Göring in Nuremberg, 1946.
11 Josef Ganz, "Die Ausstellung der Ausstellungen: Berlin 1933," in *Motor-Kritik 13*, no. 5, early March 1933.
12 "Brief aus dem Leserkreise," in *Motor-Kritik 13*, no. 5, early March 1933.
13 Josef Ganz, "How I Invented the Volkswagen," in *Australian Motor Sports & Automobiles*, June 1965, pp. 6-10.
14 W.E. Fauner, "Wirklich Neues! Rückblick auf die Berliner Autoschau," in *Der Kraftzug in Wirtschaft und Heer* 8, no. 3, March 1, 1933, pp. 41-42.
15 "Fahrer durch die Internationale Automobil- und Motorrad-Ausstellung," in *Motor und Sport 10*, no. 12, March 19, 1933, pp. 28-29.
16 Werner Fuess, "Motor und Sport—Prüfungsbericht No. 3 Standard-'Superior,'" in *Motor und Sport 10*, no. 16, April 16, 1933, pp. 25-26.
17 Opening editorial in *Motor-Kritik 13*, no. 7, early April 1933, p. 169.
18 Reichstagsbrandverordnung, February 28, 1933.
19 Ermächtigungsgesetz, March 23, 1933.
20 Josef Ganz, "Ein Denkmal für den Ersterfinder des Benzinautomobils," in *Motor-Kritik 12*, no. 22, mid-November 1932, pp. 21-22.
21 Josef Ganz, "Die Hülle fiel," and Walter Ostwald, "Das Mannheimer Benz-Denkmal," in *Motor-Kritik 13*, no. 9, early May 1933, pp. 217-218 & p. 218.

Chapter Ten: Gestapo Cellar on the Alexanderplatz

1 Letter from Josef Ganz to Max J. B. Rauck, custodian of the Deutsches Museum in Munich, February 5, 1962. [Corporate History Department, Volkswagen AG, Wolfsburg]
2 Frank Arnau, *Gelebt, geliebt, gehasst: Ein Leben im 20. Jahrhundert*, K. Desch, Munich, 1972.
3 *Karteikarten*, by Frank Arnau and Paul Georg Ehrhardt, 1933. [Hessisches Hauptstaatsarchiv, Wiesbaden]
4 'Frank Arnaus neues Buch: Die braune Pest," *Wiener Sonn- und Montags-Zeitung*, October 1933.
5 Police report, Hans Panskus, April 27, 1933. [Bundesarchiv, Berlin]
6 Letter from the directors of BMW to Josef Ganz, May 3, 1933. [Hessisches Hauptstaatsarchiv, Wiesbaden]
7 Letter from Josef Ganz to Max J. B. Rauck, custodian of the Deutsches Museum in Munich, February 5, 1962. [Corporate History Department, Volkswagen AG, Wolfsburg]
8 Josef Ganz, "Zehn Jahre," in *Motor-Kritik 13*, no. 9, early May 1933, pp. 219-221.
9 Autobiography of Jakob Werlin. [manuscript archives, Mercedes-Benz Archives & Collections, Daimler AG, Stuttgart]
10 Letter from Franz Josef Popp of BMW to Jakob Werlin, June 27, 1933. [Mercedes-Benz Archives & Collections, Daimler AG, Stuttgart]
11 Patent AT 147521 ("Stromlinienförmiger, die Räder auch an den Außenseiten umschließender Wagenkasten für vier- und mahrrädrige Kraftfahrzeuge"), property of Josef Ganz, application dated May 12, 1933, granted on June 15, 1936.
12 Josef Ganz, "Für den Wagen des kleinen Mannes," in *Motor-Kritik 13*, no. 10, mid-May 1933, pp. 246-257 & 260-264.
13 Josef Ganz, "How I Invented the Volkswagen," in *Australian Motor Sports & Automobiles*, June 1965, pp. 6-10.
14 Letter from Josef Ganz to Dipl. Ing. Max J. B. Rauck, curator of the Deutsches Museum in Munich, February 5, 1962. [Corporate History Department, Volkswagen AG, Wolfsburg]
15 "Kaleidoskop" column, in *Motor-Kritik 13*, no. 11, early June 1933, p. 291.
16 Josef Ganz, "How I Invented the Volkswagen," in *Australian Motor Sports & Automobiles*, June 1965, pp. 6-10.
17 Letter from Josef Ganz to Max J. B. Rauck, custodian of the Deutsches Museum in Munich, February 5, 1962. [Corporate History Department, Volkswagen AG, Wolfsburg]
18 "Berliner Skandall, der nach Frankfurt übergreiff / Frank Arnau—alias Heinrich Schmitt—ein vielseitiges Schwindelgenie wird gesucht," in *Frankfurter Zeitung*, May 1933.
19 Letter from Josef Ganz to Max J. B. Rauck, custodian of the Deutsches Museum in Munich, February 5, 1962. [Corporate History Department, Volkswagen AG, Wolfsburg]

20 Letter from Das Verhöramt des Kantons Graubünden to H. Bechhold Verlagsbuchhandlung, June 9, 1933. [author's archive]
21 In *Motor-Kritik 13*, no. 12, mid-June 1933.
22 Georg Ising, "Die verflixte Kritik," in *Motor-Kritik 13*, no. 12, mid-June 1933, pp. 297-298.
23 Letter from Josef Ganz to Max J. B. Rauck, custodian of the Deutsches Museum in Munich, February 5, 1962. [Corporate History Department, Volkswagen AG, Wolfsburg]
24 Ibid.
25 Letter from H. Bechhold Verlagsbuchhandlung to Josef Ganz, June 29, 1933. [Hessisches Hauptstaatsarchiv, Wiesbaden]
26 Josef Ganz, "Liebe 'M.-K.'-Freunde!," in *Motor-Kritik 13*, no. 13, early July 1933.
27 Letter from Ferdinand Fuchs to the district court in Zürich, June 9, 1944. [author's archive]

Chapter Eleven: Volkswagens for the Autobahn

1 "Kaleidoskop. Allen Schädigern zum Trotz: Fünf Weltrekorde auf Standard-Superior, Bauart Ganz (Mercedes)," in *Motor-Kritik 13*, no. 14, mid-July 1933.
2 Reinhard von König-Fachsenfeld, "Warum Rekordversuche?," in *Motor-Kritik* 13, no. 15, early August 1933, p. 372-374.
3 "Kaleidoskop. Allen Schädigern zum Trotz: Fünf Weltrekorde auf Standard-Superior, Bauart Ganz (Mercedes)," in *Motor-Kritik 13*, no. 14, mid-July 1933.
4 Anon., "Das jüngste Kind—500-ccm-Standard-Superior," in *Allgemeine Automobilzeitung*, no. 8, 1933, p. 11.
5 Wolfgang von Lengerke, "Motor und Sport—Prüfungsbericht A No.163 Standard-'Superior,'" in *Motor und Sport 10*, no. 30, July 23, 1933, pp. 13-14.
6 "Standard-Superior," in *ADAC-Motorwelt 30*, no. 29, July 21, 1933, pp. 7-8.
7 *Motor-Kritik 13*, no. 15, early August 1933.
8 *Motor-Kritik 13*, no. 15, early August 1933, p. 370.
9 Standard Fahrzeugfabrik pamphlet, July 1933.
10 Josef Ganz, "Ein Wochenende," in *Motor-Kritik 13*, no. 16, mid-August 1933, pp. 395-397.
11 "Schafft Wettbewerbe für das Kleinstauto!," in *Motor*, no. 6, June 1933, p. 5.
12 Josef Ganz, "Von kleinsten Fahrzeugen und großen Leistungen. Der erste Kleinstfahrzeugbewerb des ADAC, 15.–17. August 1933," in *Motor-Kritik 13*, no. 17, early September 1933, pp. 422-437.
13 Folder "Sieg an Sieg reiht der Standard-Superior Stromlinien-Kleinwagen" Standard Fahrzeugfabrik GmbH, Ludwigsburg, 1933.
14 Felix Korn, "Standard-Superior Bergerprobung," in *Motor-Kritik 13*, no. 17, early September 1933, pp. 419-421.
15 Letter from H. I. Hoffmann in Munich to Josef Ganz in Frankfurt, September 28, 1933. [author's archive]
16 "Vom deutschen Kleinwagenmarkt" in *Motor und Sport 10*, no. 42, 1933, p. 7.
17 *Motor-Kritik* no. 19, early October 1933.
18 Letter from the editors of *Die Autobahn und "Die Reichsautobahn"—das Fachblatt für das Autobahnwesen* to Josef Ganz, November 21, 1933. [author's archive]
19 Josef Ganz, "Autobahnen befruchten die Kraftfahrzeugindustrie," in *Die Autobahn und "Die Reichsautobahn"—das Fachblatt für das Autobahnwesen*, 1934.
20 Karl Schreiner, "Den neuen Straßen neue Autos! Das Schnellverkehrsauto der Reichsautobahnen muβ erst geboren werden!," in *Automobilia*, September 1933, p. 6.
21 Letter from Walter Ostwald to Josef Ganz, October 23, 1933. [author's archive]
22 Georg Ising, "Könner im Dunkel," in *Motor-Kritik 13*, no. 20, mid-October 1933, pp. 491-493.
23 Letter from Josef Ganz to Max J. B. Rauck, custodian of the Deutsches Museum in Munich, February 5, 1962. [Corporate History Department, Volkswagen AG, Wolfsburg]

Chapter Twelve: Hitler Sketches the Beetle

1 Nachtrag zur Niederschrift der Fabrikantensitzung betr. Volkswagen, May 28, 1934. [Mercedes-Benz Archives & Collections, Daimler AG, Stuttgart]
2 "Der Weg zum deutschen Volkswagen!," in *Motor*, no. 12, December 1933, pp. 23-24.
3 Notes to Griffith Borgeson, editor of *Automobile Quarterly*, re: interview with Hans Rumpler, son of Edmund Rumpler, February 18, 1980. [author's archive]
4 Autobiography of Jakob Werlin. [manuscript archives, Mercedes-Benz Archives & Collections, Daimler AG, Stuttgart]
5 Letter from the directors of Daimler-Benz to Josef Ganz, November 19, 1933. [author's archive]
6 Agreement between Josef Ganz and Daimler Benz AG, January 22, 1935. [Hessisches Hauptstaatsarchiv, Wiesbaden]
7 Letter from the directors of Daimler-Benz to Josef Ganz, January 12, 1934. [author's archive]
8 "Kaleidoskop" column, in *Motor-Kritik 14*, no. 2, mid-January 1934.
9 Verdict of the district court in Berlin, December 4, 1933. [author's archive]
10 "Kaleidoskop" column, in *Motor-Kritik 13*, no. 24, mid-December 1933.
11 Patent DRP [DE] 549602 ("Dreirädriges Kraftfahrzeug"), property of Tatra, application dated September 15, 1929, granted on April 14, 1932.
12 Patent DRP [DE] 587409 ("Antriebsblock für Kraftfahrzeuge"), property of Josef Ganz, application dated May 8, 1932, granted on October 19, 1933.
13 Letter from Josef Ganz to Béla Barényí, October 3, 1933 [author's archive]
14 Sketch by Adolf Hitler, December 23, 1933. [author's archive]
15 K. B. Hopfinger, *Beyond Expectation: The Volkswa-*

gen Story, G. T. Foulis, London, 1954.

16 Josef Ganz, "MB 130," in *Motor-Kritik 14*, no. 1, early January 1934, pp. 6-10.

17 Letter from H. Bechhold Verlagsbuchhandlung to Josef Ganz, December 30, 1933. [Hessisches Hauptstaatsarchiv, Wiesbaden]

18 "1934," in *Motor-Kritik 14*, no. 1, early January 1934.

19 Josef Ganz, "Als Pfadfinder auf dem Wege vom Benz-Komfortable zum MB 130 Heckwagen. Ein stuck technischer Geschichte, an dem ich teilhatte," in *Motor-Kritik 14*, no. 1, early January 1934, pp. 3-5.

20 From a statement to the court by Georg Ising, February 1934. [author's archive]

21 Ibid.

22 *Exposé betreffend den Bau eines deutschen Volkswagens* by Ferdinand Porsche, January 17, 1934. [author's archive]

23 Georg Ising, "Deutsche Rennwagen 1934," in *Motor-Kritik 14*, no. 3, early February 1934, pp. 60-61.

24 Minutes of a meeting about the "Kleinstkraftwagen," February 12, 1934. [Mercedes-Benz Archives & Collections, Daimler AG, Stuttgart]

25 Letter from the RDA to the Minister of Transportation, February 19, 1934. [Mercedes-Benz Archives & Collections, Daimler AG, Stuttgart]

26 Josef Ganz, "M-K Zeitsorgen," in *Motor-Kritik 14*, no. 5, early March 1934, pp. 97-99.

Chapter Thirteen: Kaiserhof Hotel, Berlin

1 Josef Ganz, "Volkskanzler—Volksfahrzeug—Volksausstellung. Auftakt," in *Motor-Kritik 14*, no. 6, mid-March 1934, pp. 129-134.

2 Brochure describing the Tatra 77, 1934.

3 Ivan Margolius and John G. Henry, *Tatra—The Legacy of Hans Ledwinka*, SAF Publishing Ltd, Harrow, 1990.

4 Josef Ganz, "Übersicht," in *Motor-Kritik 14*, no. 6, mid-March 1934, pp. 135-148 & 150.

5 *Automobiltechnische Zeitschrift (ATZ)* no. 4, 1934.

6 *Motor*, March 1934.

7 Wolfgang von Lengerke, "Die Karosserie—der Kampf um Form," in *Motor und Sport 11*, no. 12, March 25, 1934, pp. 9-10.

8 Wolfgang von Lengerke, "Alles wie bei uns—und doch anders in Amerika," in *Kraftverkehrs-Pressedienst—Pressedienst des Gesamten Kraftverkehrswesens*, no. 7, February 14, 1934.

9 German newsreel, March 1934.

10 Josef Ganz, "Übersicht," in *Motor-Kritik 14*, no. 6, mid-March 1934, pp. 135-148 & 150.

11 Ibid.

12 Ibid.

13 Autobiography of Jakob Werlin. [manuscript archives, Mercedes-Benz Archives & Collections, Daimler AG, Stuttgart]

14 Note by Jakob Werlin on a sketch by Adolf Hitler after their meeting with Ferdinand Porsche in the Kaiserhof Hotel in Berlin, 1934. [author's archive]

15 Letter from Georg Ising to the military government in Munich, November 1, 1945. [author's archive]

16 Letter from VDO Tachometer to Josef Ganz, March 23, 1934. [Hessisches Hauptstaatsarchiv, Wiesbaden]

17 Minutes of a meeting about the development of the Volkswagen, April 11, 1934. [Mercedes-Benz Archives & Collections, Daimler AG, Stuttgart]

18 "Observation" about the development of the Volkswagen, May 5, 1934. [Bundesarchiv Berlin]

19 Veit Harlan, *Im Schatten meiner Filme*, Sigbert Mohn Verlag, Gütersloh, 1966.

20 Letter from Josef Ganz to Max J. B. Rauck, custodian of the Deutsches Museum in Munich, February 5, 1962. [Corporate History Department, Volkswagen AG, Wolfsburg]

21 Statement to the court by Paul G. Ehrhardt, April 23, 1934. [author's archive]

22 Statement to the court by Paul G. Ehrhardt, April 25, 1934. [author's archive]

23 Statement to the court by Hans Gustav Röhr, May 26, 1934. [author's archive]

24 Synopsis and Index Card on Paul G. Ehrhardt, American occupying forces, February 12, 1947. [Bundesarchiv Berlin]

25 Verdict of the Patent Office in Berlin, March 29, 1934. [author's archive]

26 Verdict of the Patent Office in Berlin, May 9, 1934. [author's archive]

27 Patent DRP [DE] 587409 ("Antriebsblock für Kraftfahrzeuge"), property of Josef Ganz, application dated May 8, 1932, granted on October 19, 1933.

28 Patent DRP [DE] 469644 ("Anordnung des Motors von Motorfahrzeugen an dem vorderen Ende eines in der Längsmittelebene des Fahrzeugs befindlichen Tragkörpers"), property of Tatra, application dated March 30, 1926, granted on December 6, 1928.

29 Design drawing by Paul Jaray for streamlined bodywork on the chassis of the Bungartz Butz, May 9, 1934. [ETH-Bibliothek, Zürich]

30 Nachtrag zur Niederschrift der Fabrikantensitzung betr. Volkswagen, May 28, 1934. [Mercedes-Benz Archives & Collections, Daimler AG, Stuttgart]

31 Letter from Franz Josef Popp of BMW to Robert Allmers of the RDA, June 1, 1934. [Mercedes-Benz Archives & Collections, Daimler AG, Stuttgart]

32 Letter from Franz Josef Popp of BMW to Wilhelm Kissel of Daimler-Benz, June 5, 1934. [Mercedes-Benz Archives & Collections, Daimler AG, Stuttgart]

33 Nachtrag zur Niederschrift der Fabrikantensitzung betr. Volkswagen, May 28, 1934. [Mercedes-Benz Archives & Collections, Daimler AG, Stuttgart]

34 Letter with contract proposal from RDA to Wilhelm Kissel of Daimler-Benz, June 9, 1934. [Mercedes-Benz Archives & Collections, Daimler AG, Stuttgart]

Chapter Fourteen: Night of the Long Knives

1 Design drawing by Paul Jaray for streamlined body-

work on the chassis of the Bungartz Butz, May 9, 1934. [ETH-Bibliothek, Zürich]

2 Letter from Josef Ganz to Max J. B. Rauck, custodian of the Deutsches Museum in Munich, February 5, 1962. [Corporate History Department, Volkswagen AG, Wolfsburg]

3 "Klausen 1934. Deutsche Rennwagen in Front," in *Motor-Kritik 14*, no. 16, mid-August 1934.

4 Horst Mönnich, *Die Autostadt*, Munich, 1951.

5 "Alpenfahrt—Etappenziel in St. Moritz," in *Motor-Kritik 14*, no. 17, early September 1934.

6 "Gesetz über die Gewährung von Straffreiheit," August 7, 1934.

7 Verdict of the court in Frankfurt in the case of Josef Ganz vs. Paul Ehrhardt, Gustav Röhr, and Bernhard Stoewer, February 4, 1935. [author's archive]

8 Letter from Josef Ganz to Max J. B. Rauck, custodian of the Deutsches Museum in Munich, February 5, 1962. [Corporate History Department, Volkswagen AG, Wolfsburg]

9 Letter from Georg Ising in Frankfurt to Josef Ganz, September 25, 1934. [author's archive]

10 Letter from *Der Motorist Presse- und Nachrichtendienst* to the German editorial office, August 23, 1934. [Mercedes-Benz Archives & Collections, Daimler AG, Stuttgart]

11 Letter from the RDA to the manufacturers of passenger cars among its members, October 13, 1934. [Mercedes-Benz Archives & Collections, Daimler AG, Stuttgart]

12 Brochure describing the Standard Superior, 1934.

13 Statement concerning the conversation between Josef Ganz and Herr Goldstein, former chairman of the Reichsverbandes der Kraftfahrzeugteile-Industrie (RKI), signed by Josef Ganz and Alfred Rüdinger, October 15, 1934. [author's archive]

14 Statement by Philip Harry Cummings, December 2, 1934. [Hessisches Hauptstaatsarchiv, Wiesbaden]

15 Otto Fischer, *Eine Antwort auf die Greuel- und Boykotthetze der Juden im Ausland*, Frankfurt a.M., December 1934.

16 Amstvermerk of the Sicherheitskorps Vaduz in Liechtenstein, December 10, 1934. [Landesarchiv Fürstentum Liechtenstein]

17 Letter from Josef Ganz to the Fürstliche Regierung von Liechtenstein, December 17, 1934. [Landesarchiv Fürstentum Liechtenstein]

18 Passport of the Republic of Honduras, Josef Ganz, January 19, 1935. [author's archive]

19 Decision of the court in Frankfurt in the case of Josef Ganz vs. Paul Ehrhardt, Gustav Röhr, and Bernhard Stoewer, February 4, 1935. [author's archive]

20 Statement to the court by Johann Martin Danner, March 5, 1935. [author's archive]

21 Statement by Johann Martin Danner about his political activities in Germany, March 1, 1935. [author's archive]

22 Statement to the court by Johann Martin Danner, March 4, 1935. [author's archive]

23 Letter from Josef Ganz to Max J. B. Rauck, custodian of the Deutsches Museum in Munich, February 5, 1962. [author's archive]

24 Statement to the court by Johann Martin Danner, March 5, 1935. [author's archive]

25 Letter from Josef Ganz to the Polizeikommando des Kantons St. Gallen, March 23, 1935. [author's archive]

26 Anon., "Christl-Marie Schultes—erfolgreiche Pilotin oder Glücksritterin mit Gröβenwahn?," in *VPD-Nachrichten*, no. 2, 2005 .

27 Letter from Josef Ganz to the Polizeikommando des Kantons St. Gallen, March 23, 1935. [author's archive]

28 Letter from the Präsident der Reichsschrifttumskammer in Berlin to Josef Ganz, March 7, 1935. [Hessisches Hauptstaatsarchiv, Wiesbaden]

29 Statement by lawyer Heinrich Kuntze of Daimler-Benz, April 7, 1935. [author's archive]

30 Letter from Josef Ganz to Theodor Hartherz, June 12, 1935. [author's archive]

31 Letter from Theodor Hartherz to Josef Ganz, June 14, 1935. [author's archive]

32 Letter from Josef Ganz to the Polizeiamt der Stadt Zürich, October 27, 1935. [author's archive]

33 "Deutschland, Deutschland über alles! Ein Rennwagen soll Europa erobern," in *Berner Tagwacht*, July 22, 1935.

34 Letter from the NSKK-Korpsführung Stabsabteilung to Korpsführer Adolf Hühnlein of the NSKK, September 5, 1935. [Mercedes-Benz Archives & Collections, Daimler AG, Stuttgart]

35 Letter from Jakob Werlin of Daimler-Benz AG in Munich to director Erich Friedrich Muff of Mercedes-Benz-Automobil AG in Zürich, November 12, 1935. [Mercedes-Benz Archives & Collections, Daimler AG, Stuttgart]

36 Letter from director Erich Friedrich Muff of Mercedes-Benz-Automobil AG in Zürich to Jakob Werlin of Daimler-Benz AG in Munich, December 4, 1935. [Mercedes-Benz Archives & Collections, Daimler AG, Stuttgart]

37 Letter from director Erich Friedrich Muff of Mercedes-Benz-Automobil AG in Zürich to Jakob Werlin of Daimler-Benz AG in Munich, April 6, 1936. [Mercedes-Benz Archives & Collections, Daimler AG, Stuttgart]

Chapter Fifteen: Swiss Volkswagen

1 Letter from Ferdinand Fuchs to the district court in Zürich, June 9, 1944. [author's archive]

2 Letter from the Städtisches Arbeitsamt Zürich to Vorstand des Gesundheitsamtes Zürich, January 15, 1936. [Schweizerisches Bundesarchiv, Bern]

3 Letter from Ferdinand Fuchs to the Polizeiinspektorat der Stadt Zürich, October 24, 1935. [Schweizerisches Bundesarchiv, Bern]

4 Letter from Josef Ganz to Herr Hüni of the Polizeiamt der Stadt Zürich, October 27, 1935. [author's archive]
5 Letter from Städtisches Arbeitsamt Zürich to the board of the Gesundheitsamt Zürich, January 15, 1936. [Schweizerisches Bundesarchiv, Bern]
6 Letter from Städtisches Arbeitsamt Zürich to the board of the Gesundheitsamt Zürich, June 29, 1936. [Schweizerisches Bundesarchiv, Bern]
7 Minutes of a meeting about Porsche's prototypes at Daimler-Benz in Berlin, February 24, 1936. [Corporate History Department, Volkswagen AG, Wolfsburg]
8 Notes of a meeting with Ghislaine Kaes, Ferdinand Porsche's former secretary, November 25, 1963. [Corporate History Department, Volkswagen AG, Wolfsburg]
9 Postcard from Georg Ising in Frankfurt to Josef Ganz in Zürich, February 16, 1936. [author's archive]
10 Statement by Josef Ganz to the Police de Sureté du Canton de Vaud, Lausanne, March 30, 1936. [Schweizerisches Bundesarchiv, Bern]
11 Letter from Josef Ganz to Max J. B. Rauck, custodian of the Deutsches Museum in Munich, March 2, 1962. [author's archive]
12 Report by lawyer C. Wiegand of the high court in Berlin, May 9, 1936. [author's archive]
13 Verdict of the National Patent Office in the case of Tatra vs. Wilhelm Gutbrod of Standard Fahrzeugfabrik GmbH and Josef Ganz, June 27, 1936. [author's archive]
14 Letter from Standard Fahrzeugfabrik GmbH to lawyer Richard Rheinstein, July 22, 1935. [author's archive]
15 Statement by Wilhelm Gutbrod, June 17, 1941. [Hessisches Hauptstaatsarchiv, Wiesbaden]
16 Walter Krumnow, "Vom Schweren Brocken in die leichte Wanze," in *Das Motorrad*, no. 29, July 18, 1936, pp. 1223-1226.
17 Autobiography of Jakob Werlin. [manuscript archives, Mercedes-Benz Archives & Collections, Daimler AG, Stuttgart]
18 Letter from Standard Fahrzeugfabrik GmbH to Otto Schirz of the Reichsverband der Automobilindustrie (RDA), September 2, 1936. [author's archive]
19 Letter from Dr. Ing. h.c. F. Porsche GmbH to the Reichsverband der Automobilindustrie (RDA), October 5, 1936. [author's archive]
20 Letter from Dr. Ing. h.c. F. Porsche GmbH to the Reichsverband der Automobilindustrie (RDA), November 3, 1936. [author's archive]
21 Verdict of the National Patent Office in the case of Tatra vs. Wilhelm Gutbrod, Standard Fahrzeugfabrik and Josef Ganz, October 31, 1936. [author's archive]
22 Letter from Fritz Isler to Ferdinand Fuchs, November 10, 1936. [Hessisches Hauptstaatsarchiv, Wiesbaden]
23 Letter from Georg Schlesinger of the University of Brussels to Ferdinand Fuchs, April 15, 1936. [author's archive]
24 Patent CH 200876 ("Kraftwagen"), property of Erfindungs- und Finanzierungs AG (ERFIAG), application dated October 26, 1937, granted on October 31, 1938.
25 Patent CH 200708 ("Mitrailleur-Geländewagen") property of Erfindungs- und Finanzierungs AG (ERFIAG), application dated July 8, 1937, granted on October 31, 1938.
26 Letter from Josef Ganz to Eugen Fontanellaz of the Bundesamt für Industrie, Gewerbe und Arbeit (BIGA), November 6, 1938. [Schweizerisches Bundesarchiv, Bern]
27 Anon., "Ein Schweizer Volkswagen?," in *Automobil-Revue*, no. 99, December 10, 1937.
28 *Touring*, no. 3, January 20, 1938.
29 Georg Ising, "Eine schweizer Kleinwagenkonstruktion," in *Motor-Kritik 18*, no. 1, early January 1938.
30 Letter from Rapid Motormäher AG to the Direktion der Volkswirtschaft des Kantons Zürich, November 10, 1938. [Schweizerisches Bundesarchiv, Bern]
31 Patent DRP [DE] 642192 ("Kraftfahrzeug mit vier unabhängig abgefederten Rädern"), property of Daimler-Benz, application dated April 22, 1932, granted on February 11, 1937.
32 Letter from lawyer Fritz Isler of Zürich to ERFIAG, July 15, 1937. [Hessisches Hauptstaatsarchiv, Wiesbaden]
33 Agreement between Daimler-Benz and ERFIAG, August 15, 1938. [Hessisches Hauptstaatsarchiv, Wiesbaden]
34 "Bekanntmachung," in *Deutschen Reichsanzeiger und Preußischer Staatsanzeiger*, no. 187, August 13, 1938.
35 Verdict of the district court in Berlin in the case of Ambi-Budd Presswerke vs. ERFIAG, September 9, 1938. [Hessisches Hauptstaatsarchiv, Wiesbaden]
36 Verdict of the district court in Berlin in the case of Ambi-Budd Presswerke vs. ERFIAG, July 8, 1939. [Hessisches Hauptstaatsarchiv, Wiesbaden]
37 Press release Der Leichtwagen, System GANZ. [author's archive]
38 Letter from Erich Skischally of Volkswagen to Ferdinand Porsche, June 9, 1939. [author's archive]
39 Letter from Paul G. Ehrhardt to the Städtische Fremdenpolizei, July 14, 1939. [author's archive]
40 Letter from Ferdinand Fuchs to the district court in Zürich, June 9, 1944. [author's archive]
41 Anon., "Vom Ungeist eines Polizeifunktionärs," in *Freies Volk*, January 30, 1948.
42 Letter from Wälchli of the Bundesamt für Industrie, Gewerbe und Arbeit (BIGA) to Eidgenöstischen Fremdenpolizei, February 25, 1941. [author's archive]
43 Einweisung in ein Arbeitslager für Emigranten, Eidgenössisches Justiz- und Polizeidepartment, June 17, 1941. [author's archive]

44 Patent CH 214684 ("Schmiereinrichtung für das untere Pleuellager von Kolbenmaschinen mit Sprühölschmierung"), property of Erfindungs- und Finanzierungs AG (ERFIAG), application dated June 27, 1940, granted on May 15, 1941.
45 Patent CH 270614 ("Aufnahmevorrichtung für zu schleifende Spiralbohrer"), property of Erfindungs- und Finanzierungs AG (ERFIAG), application dated April 18, 1945, granted on September 15, 1950.
46 Patent CH 230011 ("Vorrichtung zur Angleichung des Antriebskurbel-Rundlaufs an eine, geringere Muskelermüdung hervorrufende Bewegungskurve"), property of Karl Fleischmann and Madeleine Paqué, application dated September 5, 1941, granted on November 30, 1943.
47 Patent CH 252935 ("Abfederungseinrichtung für Fahrzeugräder, insbesondere für Veloräder"), property of Josef Ganz, application dated June 28, 1944, granted on January 31, 1948.
48 Verdict of the district court in Berlin in the case of Tatra vs. Josef Ganz, June 14, 1941. [author's archive]
49 Statement by Wilhelm Gutbrod, June 17, 1941. [Hessisches Hauptstaatsarchiv, Wiesbaden]
50 Notes to Griffith Borgeson, editor of *Automobile Quarterly*, re: interview with Hans Rumpler, son of Edmund Rumpler, February 18, 1980. [author's archive]
51 Report entitled Selbstkosten—Berechnung für den GANZ-WAGEN, Hispano-Suiza SA, April 2, 1942. [author's archive]
52 Police report by Otto Dätwyler and Otto Gloor, January 18, 1944. [author's archive]
53 Anon., "Merkwürdige Sprünge des Amtsschimmels," in *National-Zeitung*, June 29, 1944.
54 Anon., "Vor Bezirksgericht 'Interniert den Parasiten!,'" in *Tages-Anzeiger für Stadt und Kanton Zürich*, June 29, 1944.
55 Article in four parts, entitled "Fremdenpolizei in Dienst der Gestapo?," in *Volksrecht*, January 29, January 30, February 1, and February 6, 1945.

Chapter Sixteen: Sanctuary in Australia

1 Jonathan E. Helmreich, "The Bombing of Zürich," in *Aerospace Power Journal*, summer 2000.
2 Statement by Alfred Bloch, employee of the Economic Intelligence Division United States Group Control-Council, September 1, 1945. [author's archive]
3 Biographical report on Jakob Werlin by the Research and Analysis Branch. [Donovan Nuremberg Trial Collection, Cornell University Law Library]
4 Interrogation of Georg Eidenschink by Capt. O. N. Nordon, November 6, 1945. [Donovan Nuremberg Trial Collection, Cornell University Law Library]
5 Letter from Georg Ising to the military government in Munich, November 1, 1945. [author's archive]
6 Letter from "ER" to Josef Ganz, June 10, 1946. [archive of Dieter Klüpfel]
7 Letter from Leo Widmer to the Bundesamt für Industrie, Gewerbe und Arbeit (BIGA), July 28, 1944. [Schweizerisches Bundesarchiv, Bern]
8 Letter from Arnold Rütishauser to Erfindungs- und Finanzierungs AG (ERFIAG), March 18, 1944. [author's archive]
9 Georg Ising, "Stromlinie und Volkswagen," in *Motor Rundschau 1*, no. 5, March 1, 1947, p. 65.
10 Georg Ising, "Der verhinderte Volkswagen," in *Motor Rundschau 1*, no. 5, March 1, 1947, pp. 66-67.
11 Frank A. E. Martin, "Der GANZ-Kleinstwagen," in *Das Auto 4*, 1, January 1949, pp. 6-7 & 22.
12 Letter from Societá Commerciale Aeronautica MACCHI to Ferdinand Fuchs, June 25, 1948. [Schweizerisches Bundesarchiv, Bern]
13 Anon., "Der helvetische Besen," in *Sie und Er*, February 20, 1948.
14 Anon., "Vom Ungeist eines Polizeifunktionärs," in *Freies Volk*, January 30, 1948.
15 Letter from Polizeikommando des Kantons Zürich to Polizeikommando Zürich, January 27, 1948. [author's archive]
16 Letter from lawyers Alois Grendelmeier and Walter Baechi to the district court in Zürich, March 4, 1950. [author's archive]
17 Anon., "Strafprozeβ gegen Ganz—ohne Ganz," in *Die Tat*, June 23, 1950.
18 Letter from Josef Ganz to Zentralmeldeamt der Militärregierung Rückerstattungsansprüche. [Hessisches Hauptstaatsarchiv, Wiesbaden]
19 Letter from Josef Ganz to the board of ERFIAG, November 3, 1950. [author's archive]
20 Letter from the Fremdenpolizei in Zürich to the Eidg. Justiz- und Polizeidepartement in Bern, September 9, 1950. [Schweizerisches Bundesarchiv, Bern]
21 Anon., "Ein Ehrabschneider wird verurteilt—Der Emigrant Josef Ganz gewinnt einen Ehrverletzungsprozess," in *National-Zeitung*, October 13, 1950.
22 Statement by Z. Bloch-Moreag of Dimokratikus, November 19, 1950. [Schweizerisches Bundesarchiv, Bern]
23 *Neuen Welt*, 1963.
24 Anon., "Ein Ehrabschneider wird verurteilt—Der Emigrant Josef Ganz gewinnt einen Ehrverletzungsprozess," in *National-Zeitung*, October 13, 1950.
25 Letter from Margit Ganz to Josef Ganz, November 8, 1950. [author's archive]
26 Interview with Gerry Harant, one-time colleague and friend of Josef Ganz in Australia, July 2008. [author's archive]
27 General Motors—Holden's Ltd. Engineering Department Organization Chart, ca. 1955. [author's archive]
28 John Mulliken, "Volkswagen 'A Revelation of a Secret Love,'" in *LIFE International*, October 24, 1960, pp. 73-74.
29 Agreement between Josef Ganz and the German state of Hesse, January 22, 1959. [Hessisches Hauptstaatsarchiv, Wiesbaden]

30 Letter from Heinrich Nordhoff of Volkswagen to Josef Ganz, February 28, 1961. [Corporate History Department, Volkswagen AG, Wolfsburg]
31 Letter from Josef Ganz to Heinrich Nordhoff of Volkswagen, March 9, 1961. [Corporate History Department, Volkswagen AG, Wolfsburg]
32 Letter from Josef Ganz to Heinrich Nordhoff of Volkswagen, June 21, 1961. [Corporate History Department, Volkswagen AG, Wolfsburg]
33 Letter from Heinrich Nordhoff of Volkswagen to Josef Ganz, July 17, 1961. [Corporate History Department, Volkswagen AG, Wolfsburg]
34 Letter from Josef Ganz to Heinrich Nordhoff of Volkswagen, August 9, 1961. [Corporate History Department, Volkswagen AG, Wolfsburg]
35 Letter from Josef Ganz to Heinrich Nordhoff of Volkswagen, October 27, 1961. [Corporate History Department, Volkswagen AG, Wolfsburg]
36 Patent DRP [DE] 636633 ("Lagerung eines Antriebsblockes an Kraftfahrzeugen, insbesondere mit einem z. B. rohrfoermigen Mittelrahmen"), property of Tatra, application dated December 18, 1932, granted on January 7, 1937.
37 Dieter Großherr, "Das Wunder auf Bestellung—Vor 25 Jahren: die ersten Volkswagen—Ein Auto wird zum Politikum," in *Frankfurter Rundschau*, October 28, 1961, p. 48.
38 Dieter Großherr, "Der Volkswagen feiert seinen 25 Geburtstag—Die Entwicklung begann schon um 1930 / Zwei heute fast vergessene Männer standen Pate," in *Suddeutsche Zeitung*, October 20, 1961.
39 Dieter Großherr, "Die VW Story," in *Gute Fahrt*, no. 9, 1961, pp. 30-31, 66-69.
40 Letter from Josef Ganz to J. H. Horn of GM Holden, March 25, 1962. [author's archive]
41 Letter from Josef Ganz to J. H. Horn of GM Holden, March 21, 1962. [author's archive]
42 Letter from Josef Ganz to J. H. Horn of GM Holden, March 25, 1962. [author's archive]
43 Statement by Edwin W. Knight, December 13, 1965. [Hessisches Hauptstaatsarchiv, Wiesbaden]
44 Letter from Josef Ganz to Max J. B. Rauck, custodian of the Deutsches Museum in Munich, March 2, 1962. [author's archive]
45 Letter from Josef Ganz to H. W. Gage of GM Holden, March 30, 1962. [author's archive]
46 Letter from Josef Ganz to H. W. Gage of GM Holden, March 27, 1962. [author's archive]
47 Letter from Josef Ganz to Maja von Tolnai, September 11, 1963. [author's archive]
48 Internal memo by W. Jahnke of Volkswagen to Heinrich Nordhoff of Volkswagen, May 21, 1964. [Corporate History Department, Volkswagen AG, Wolfsburg]
49 Letter from Heinrich Nordhoff of Tatra to Josef Ganz, June 23, 1964. [Corporate History Department, Volkswagen AG, Wolfsburg]
50 Letter from Josef Ganz to Heinrich Nordhoff of Volkswagen, September 13, 1964. [Corporate History Department, Volkswagen AG, Wolfsburg]
51 Letter from Heinrich Nordhoff of Volkswagen to Josef Ganz, October 5, 1964. [Corporate History Department, Volkswagen AG, Wolfsburg]
52 Letter from Josef Ganz to Heinrich Nordhoff of Volkswagen, October 13, 1964. [Corporate History Department, Volkswagen AG, Wolfsburg]
53 Letter from A. Metzner of the secretarial office at Volkswagen to Josef Ganz, October 30, 1964. [Corporate History Department, Volkswagen AG, Wolfsburg]
54 Letter from Heinrich Nordhoff of Volkswagen to Josef Ganz, June 23, 1964. [Corporate History Department, Volkswagen AG, Wolfsburg]
55 Letter from Ernst Ganz to Josef Ganz, May 31, 1965. [author's archive]
56 Letter from "Mary" to Ernst Ganz, April 26, 1965. [author's archive]
57 Cyril Posthumus, "Whose Was the Volkswagen? Doubts on the Origin of a Well-Known Species," in *Motor*, February 27, 1965, pp. 106-107, 109 & 111.
58 Josef Ganz, "How I Invented the Volkswagen," in *Australian Motor Sports & Automobiles*, June 1965, pp. 6-10.
59 Letters from Josef Ganz to Ernest Schmid, August 10 and 13, 1964. [author's archive]
60 Letter from the Embassy of the Federal Republic of Germany to the Department of External Affairs of the Commonwealth of Australia, October 18, 1965. [National Archives of Australia]
61 Letter from the Prime Minister's Department to the Department of External Affairs, November 24, 1965. [National Archives of Australia]
62 Letter from Josef Ganz to Max J. B. Rauck, custodian of the Deutsches Museum in Munich, February 13, 1966. [author's archive]
63 Letter from Josef Ganz to Margit Ganz, November 9, 1966. [author's archive]
64 Ibid.
65 John Winding-Sorensen, "Josef Ganz: The Unsung Sire?," in *Car*, October 1972, pp. 61-63.